GREENWICH OBSERVATORY

GREENWICH OBSERVATORY

*One of three volumes by different authors
telling the story of Britain's oldest scientific institution*

THE ROYAL OBSERVATORY
AT
GREENWICH AND HERSTMONCEUX
1675 - 1975

Volume 3: The Buildings and Instruments

by

Derek Howse

NATIONAL MARITIME MUSEUM

TAYLOR & FRANCIS · LONDON

1975

First published 1975 by Taylor & Francis Ltd.,
10–14 Macklin Street, London WC2B 5NF.

© 1975 Derek Howse

ISBN 0 85066 095 5

Design and production in association with Book Production
Consultants, 125, Hills Road, Cambridge.
Set in 12/13 Monotype Bembo Series 270 by The Lancashire
Typesetting Co. Ltd., Bolton.
Printed by Taylor & Francis (Printers) Ltd., Rankine Road,
Basingstoke, Hampshire.
Bound and slip-cased by the University Printing House,
Shaftesbury Road, Cambridge.
Blocking brass design by William Andrewes.

Introduction

IT was sound common sense, if perhaps also a certain royal parsimony, that led King Charles II in 1675 to direct his Master General of the Ordnance to build Wren's observatory on Greenwich hill overlooking the royal Palace. It was necessity that prompted the tenth Astronomer Royal, more than 250 years later, to abandon the old Wren building for a castle in Sussex almost twice as old. The smoke and lights of modern London had made the Greenwich site untenable for astronomers.

Those members of the staff of the Royal Greenwich Observatory at Herstmonceux whose service goes back before World War II to the days when it was the Royal Observatory at Greenwich tend to be proud of the fact. Something of this pride rubs off on their younger colleagues, who still, after all, transmit Greenwich time to a world that measures its longitude from a zero meridian passing through Greenwich. And when the old hands retrace their steps, or the youngsters make perhaps their first pilgrimage, to the old site at Greenwich, what a transformation awaits them there! The Director and staff of the National Maritime Museum inherited the buildings and such of the old instruments as were not transported to Sussex, and they have quite literally transformed the place from a workaday astronomical observatory, more than a little rundown and grubby, into a permanent exhibition embodying the story of three centuries of British astronomy.

There could be nobody better than the Head of the Museum's Astronomy Department, himself largely responsible for this transformation, to describe in these pages the way the instruments used by successive Astronomers Royal evolved, as scientific needs changed and instrumental skills developed, and the buildings housing them changed accordingly. Derek Howse's story makes fascinating reading as a narrative, but it is also a mine of information to which future students and historians can refer.

In concentrating on instrumental matters—the astronomer gets nowhere without his instruments—this volume is a worthy companion to its two fellows which trace the history of the Royal Observatory chronologically through the three centuries of its existence.

A. HUNTER
Director, RGO

Herstmonceux Castle
January 1975

Foreword

WHEN our story begins in 1675, there was no practicable method of finding longitude at sea. For the want of this, ships were wrecked, men died, voyages lasted longer—and trade suffered.

By the middle of the seventeenth century when ocean trading voyages had become commonplace, finding a solution was one of the most urgent problems of the day. Vast money prizes were offered—by Spain, by the Netherlands, by France, and by Britain. The last—for no less than £20 000, perhaps £250 000 in today's money—was the largest prize of all and, incidentally, the only one of the four to be awarded *in toto*.

Of the many possible approaches to the longitude problem, that of astronomy seemed to offer the most promise. Indeed, theoretical solutions to the two methods which eventually did succeed—the lunar distance and the chronometer—were propounded in the sixteenth century, the first by John Werner in 1514, the second by Gemma Frisius in 1530. But it must be stressed that these solutions were only theoretical. In 1675 the practical means—accurate angle-measuring instruments, accurate marine timekeepers, and accurate astronomical tables—simply did not exist. It was to supply the last of these deficiencies that, in 1675, King Charles II was persuaded, largely by Sir Jonas Moore, to appoint the 28-year-old John Flamsteed his Astronomical Observator (a post later called Astronomer Royal) with specific directions:

"forthwith to apply himself with the utmost care and diligence to the rectifying the tables of the Motions of the Heavens and the places of the fixed stars so as to find out the so-much-desired Longitude of Places for the perfecting the Art of Navigation".*

The Royal Observatory founded by Charles II three hundred years ago still survives. The working observatory, now located at Herstmonceux in Sussex and known as the Royal Greenwich Observatory, is Britain's oldest scientific institution. Not only does it continue in "rectifying the tables of the Motions of the Heavens and the places of the fixed stars" but it also performs the many other functions in observational astronomy which fall to a national observatory. As for the old observatory buildings in Greenwich Park, these are now part of the National Maritime Museum—surely most appropriate in view of the Royal Observatory's nautical beginnings. The Old Royal Observatory is now a museum of astronomy and time: the various buildings have been severally restored to their state when in active astronomical use; instruments and clocks formerly used at Greenwich—a very high proportion of which have survived, thanks to the sense of history of the various Astronomers Royal—are now on display in their working settings, many in their original positions.

The three volumes of this work—published on the occasion of the Tercentenary of the foundation of the Royal Observatory—tell the story of Flamsteed and his successors, the first two dealing with the astronomers and their achievements, the last—the present volume— with the tools they used to do this—the instruments, the clocks, and the buildings which contained them.

The first chapter traces the development of the buildings at Greenwich, describing in broad terms the changes in equipment which gave rise to them. The next seven chapters describe and illustrate in some detail one by one the principal astronomical instruments of both Greenwich and Herstmonceux—the mural instruments, the transit instruments, the equatorial sectors, and the principal equatorial telescopes—telling how they were used, summarizing their history, and citing fuller descriptions where these have been published. The last three chapters deal in turn with the minor astronomical instruments, the magnetical and meteorological instruments, and the Observatory's clocks.

*Royal Warrant of 4 March 1675.

viii

The chief sources of information have been the published *Greenwich Observations* (which in one form or another cover the whole 300 years except for Halley's régime); the manuscript originals among the RGO papers preserved at Herstmonceux (Halley's observations can be seen here); the reports of the Board of Visitors from 1714 to 1835, copies of which are at the RGO; and the annual reports of the Astronomers Royal (published from 1836 to 1964). In general, sources of information are cited only when they fall outside the above categories.

In quotations, spelling has been modernized. Because the makers and users of most of the instruments used English feet and inches, we have done the same, the metric equivalents being given in the index.

H. D. HOWSE
Curator of Astronomy,
National Maritime Museum

November 1974

Acknowledgements

FIRST and foremost, my thanks must go to Mr Philip Laurie, archivist, and lately Head of the Solar Department at the Royal Greenwich Observatory, Herstmonceux. Over the past ten years he has unstintingly shared with me his unrivalled knowledge of the history of the buildings and instruments of the Royal Observatory and of the archives relating to them. Furthermore he has undertaken the labour of reading through the whole manuscript. I am indebted also to many of his colleagues at Herstmonceux, especially to the Director, Dr. Alan Hunter, and his predecessors for their personal help and for affording me so many facilities there, to Mr Humphry Smith, 'Mr GMT' himself, to Mr R. H. Tucker and Mr Kenneth Blackwell on transit circles, to Miss Joy Penney on the reversible transit, to Mr Michael Lowne on the equatorials, to Mr B. R. Leaton and Dr Stuart Malin on the magnetic apparatus, and to Mr David Calvert who took many of the photographs.

I am most grateful to Mr Basil Greenhill, Director, Mr David Waters, Deputy Director, and to the Trustees of the National Maritime Museum for all their help and encouragement, and for placing so many of the museum's resources at my disposal. Of my colleagues at the museum—past and present—I must make special mention of Mr Peter Ince whose many hours of laborious transcriptions have now at last been amply justified; Mr John Dix for his help on the zenith instruments; Mrs Valerie Finch for help in so many directions; Mr Brian Tremaine and his staff for the photography; and Miss Jane McAuliffe for transforming my scribbles into impeccable typescript.

Except where otherwise credited, all the illustrations are by courtesy either of the Trustees of the National Maritime Museum, or of the Director of the Royal Greenwich Observatory.

Abbreviations used in the Text

AR	Astronomer Royal
ATC	Airy Transit Circle
Dec	Declination
GMT	Greenwich Mean Time
INT	Isaac Newton telescope
MS., MSS.	Manuscript(s)
MT	Mean Time
NMM	National Maritime Museum
NPD	North Polar Distance
OG	Object-glass
ORO	Old Royal Observatory
PZT	Photographic Zenith Tube
RA	Right Ascension
RGO	Royal Greenwich Observatory
RO	Royal Observatory
RS	Royal Society
RZT	Reflex Zenith Tube
T of V	Transit of Venus
ZD	Zenith Distance

Abbreviations used in the references are listed on page 149.

List of Illustrations

xiii

Contents

1

The Buildings at Greenwich
over 300 Years

Note.—The changes in the buildings described chronologically in
this chapter are summarized building by building in Appendix I
and illustrated in Appendix III.

JOHN FLAMSTEED (b. 1646, d. 1719)
First Astronomer Royal, 1675–1719

THE circumstances surrounding the foundation of the Royal
Observatory are treated in some detail in another volume in
this series. To provide the background for this volume we will
quote from the writings of the first Astronomer Royal who is in
any case the main source of information on the subject. Alas, those
principally responsible—Sir Jonas Moore, F.R.S., Surveyor-General
of the Ordnance, and Sir Christopher Wren, the King's Surveyor-
General, sometime Professor of Astronomy in Oxford and London,
and later President of the Royal Society—have left almost no
documents on the subject. The official papers of the Board of
Ordnance—the government department responsible for footing the
bill—yield but little: the diary of Robert Hooke (1635–1703),
Secretary of the Royal Society (no friend of Flamsteed's), is almost
the only other source of value.

> "Betwixt my coming up to London, and Easter [1675], an accident
> happened that hastened, if it did not occasion the building of the
> Observatory. A Frenchman that called himself le Sieur de St. Pierre
> having some small skill in Astronomy, and made an interest with a
> French lady then in favour at Court, proposed no less than the
> Discovery of the Longitude: and had procured a kind of Commis-
> sion from the King to the Lord Brouncker, Dr Ward Bishop of
> Salisbury, Sir C Wren, Sir C Scarborow, Sir J Moore, Col Titus,
> Dr Pell, Sir Robert Murray, Mr Hook, and some other ingenious
> gentlemen about the town and Court, to receive his proposals with
> power to elect and to receive into their number any other skilful
> persons: and having heard them to give the King an account of them
> with their opinion whether or no they were practicable and would

I

show what he pretended: Sir J Moore carried me with him to one of their meetings where I was chosen into their number; and, after the Frenchman's proposals were read: which were

1. To have the year and day of the observations:
2. The height of two stars and on which side of the meridian they appeared:
3. The height of the Moon's two limbs:
4. The height of the pole.

All to degrees and minutes.

It was easy to perceive from these demands that the Sieur understood not that the best lunar tables differed from the Heavens; and that therefore his demands were not sufficient for determining the longitude of the place, where such observations were or should be made, from that to which the lunar tables were fitted: which I represented immediately to the company …"[1]

The French lady mentioned was Louise de Kéroualle, Duchess of Portsmouth, King Charles's favourite of the moment. The commission to hear St Pierre's proposals was appointed on 15 December 1674,[2] and first met on 12 February 1675. Flamsteed, having provided the observations demanded, then produced a report in which he scornfully dismissed the Sieur's claims—"upon which he huffed a little and disappeared, since which time we have heard no farther of him".[2] But, said Flamsteed, it *would* be possible to find longitude by lunar observations (though not in the way St Pierre suggested) if there were accurate star catalogues and lunar tables—but this would demand many years of observations with large instruments fitted with telescopic sights.

"When Charles II, King of England, was informed of these facts, he said the work must be carried out in royal fashion. He certainly did not want his ship-owners and sailors to be deprived of any help that the heavens could supply, whereby navigation could be made safer. Therefore his most Serene Majesty was pleased to decree that an Observatory should be built and that the necessary expenses of this building should be borne by the Office of Ordnance. Sir Jonas Moore, at that time Surveyor General of the Ordnance (it was he who had recommended me to the king as suitable for this astronomical work and, during his lifetime, on every possible occasion, he had always been most friendly towards me and helpful in all that concerned the building of the Observatory) took great trouble that nothing should be lacking in promoting and carrying through the work."[3]

As explained in the Foreword, Flamsteed was appointed 'Astronomical Observator' to the King on 4 March 1675, with the meagre salary of £100 per annum. He immediately took up residence with Sir Jonas in the Tower of London, using the north-east turret of the White Tower for his observations until a more permanent site for the Royal Observatory could be found.

> "The next thing to be thought of was a place to fix in. Several were proposed as Hyde Park and Chelsea College. I went to view the ruins of this latter and judged it might serve the turn: and the better because it was near the Court. Sir Jonas rather inclined to Hyde Park, but Sir Ch Wren mentioning Greenwich Hill it was resolved on. The King allowed 500*l* in money; with bricks from Tilbury Fort where there was a spare stock; and some wood, iron and lead from a gatehouse [Coldharbour] demolished in the Tower; and encouraged us further with a promise of affording what more should be requisite.
>
> In July following I removed from his house, where I had been kindly entertained all this summer, to Greenwich to have an eye upon the workmen. The foundation was laid Aug 10 1675 and the work carried on so well that the roof was laid and the building covered, by Christmas: of which I need give no description, because the figures of the Park, the ichnography of the house, and several prospects of it will inform the reader better than a long description."[4]

In the autumn of 1674 the Royal Society had had plans for setting up their own observatory in Chelsea College, to be equipped at Sir Jonas's expense.[5] These plans were dropped when Wren chose Greenwich Park as the place for the new Royal Observatory. The King's warrant, reprinted in full in Volume I, and dated 22 June 1675, said it should be built in Greenwich Park "upon the highest ground at or near the place where the castle stood, with lodging for our astronomical observator".

The park had been enclosed by Humphrey, Duke of Gloucester (brother of Henry V), in 1437 and a tower was built soon afterwards on the present Observatory Hill. Repaired and rebuilt by Henry VIII in 1526—he is said to have visited there "a fair lady whom he loved"—the tower was known as *Mirefleur* in Queen Elizabeth's time.[6]

According to Flamsteed, there had in Charles I's time been plans for building an observatory in the park on One Tree Hill with a meridian instrument "as large as any of the Arabs boast of".[7]

In 1642, Greenwich Castle, as it was then called, was garrisoned by Parliament. It was still standing in 1662 when a sketch of it by Wenceslaus Hollar appeared on a manuscript plan of the River Thames drawn by Jonas Moore himself (Fig. 1).

The wording of the foundation warrant just quoted, seems to indicate that the castle had been demolished by 1675. In any event, the building now called Flamsteed House was raised on the foundations of the old castle. Flamsteed, writing to his friend Richard Towneley (1629–1707), said on 22 January 1676: "It were much to be wished our walls might have been meridional but for saving of Charges it was thought fit to build upon the old ones which are some $13\frac{1}{2}°$ false and wide of the true meridian. . . ."[8]

Of the new building Wren himself said later it was for the observator's habitation and a little for pomp.[9] Its main feature was the Great Room (today's Octagon Room) seen in Fig. 2. Undoubtedly full of pomp, this room was also very functional, as one would expect from Wren, who had himself been a professor of astronomy: its tall windows allowed the long telescopes of the day to be used to the best advantage; it also accommodated Moore's two great year-clocks—with pendulums 13 feet long, hung above the movements behind the wainscot—with which Flamsteed was able to prove that, for practical purposes, the Earth rotates at a constant speed, a fact fundamental to the longitude problem but which could not be proved until after the invention of the pendulum clock in 1657, and which was not disproved until the coming of the quartz clock in the 1940s.

The real work of the Observatory was, however, carried out, not in the Great Room, but in the small building at the bottom of the garden—the Sextant House and Quadrant (later Arc) House. There Flamsteed built two meridian walls and installed his fundamental instruments.

Robert Hooke re-visited Greenwich on 30 June and 28 July 1675, and 'set out' the Observatory, presumably to Wren's designs. At 3.14 p.m. on 10 August the foundation stone was laid by Flamsteed who had moved to Greenwich to superintend the building, being given apartments in the Queen's House.

As we have seen, the roof was on by Christmas and on 22 January 1676, Flamsteed sent Towneley a description, accompanied by a detailed scale plan[10] (reproduced as Fig. 3) on which the two summer houses at the end of the north terrace—one of which was to become the solar observatory—are omitted. As the plan is otherwise so accurate, it seems most unlikely that this was a mistake on Flamsteed's part. Perhaps they were architectural afterthoughts: one can imagine Wren, standing at the bottom of the hill looking up at the new building (Fig. 4), deciding something was needed to balance the roof-turrets.

A few details emerge on the building and equipping to augment Flamsteed's own account quoted above. The roof lead came from the old Coldharbour gatehouse at the Tower;[11] ordnance stores supplied two copper balls and two great round stone shot, presumably for roof-turrets and gate-posts;[12] the Navy Board supplied spars for the long telescopes on the roof and in the garden[13] (see Fig. 5). The necessary £500 was acquired from the sale of unserviceable gunpowder from Portsmouth and the Tower to Mr Polycarpus Wharton, who presumably made it serviceable again and re-sold it to the Ordnance.[14] The total amount disbursed by Flamsteed "to several workmen and labourers for work done and materials spent and used in building of his Majesty's Observatory on Greenwich Hill within the Park between 19th November 1675 and 31st May 1677 as by a Bill of Particulars allowed and remaining in the office with several vouchers therewith annexed" amounted to £520 9s. 1d.[15]—£20 9s. 1d. overspent. Alas, the 'Bill of Particulars' mentioned has not been found among the Ordnance papers though similar accounts of disbursements between 1677 and 1682 survive in the RGO papers.[16]

The mast in the garden was erected on 1 May 1676; the first observations in the Great Room were made on 31 May. Flamsteed and his two 'servants', Thomas Smith and Cuthbert Denton, moved from the Queen's House to the Observatory on 10 July, and the first sextant observation was made on 16 September.

The Royal Observatory was complete. Flamsteed could now begin systematically to acquire his 'stock of observations' which eventually resulted, not only in the published star catalogues, but also in providing Newton with the observational data he needed to propound his gravitational theory.

Flamsteed's foundation equipment consisted of two small telescopes, the Towneley micrometer, a 3-foot quadrant and a clock—all his own property—and two year-clocks, an equatorial sextant, and Hooke's 10-foot mural quadrant—presented by Sir Jonas Moore. The Government provided buildings and salaries only—no equipment.

Lord North's presentation of the living of Burstow in Surrey in 1684, and his father's death in 1688, gave Flamsteed the means to construct a 7-foot mural arc which he mounted on the west wall of the former Quadrant House, opposite Hooke's 10-foot quadrant which had proved a failure and which by then seems to have been removed.

The 80-foot mast in the garden was taken down about 1690 when there were fears that it might be blown down onto the Arc House.[17]

A result—almost the only result—of the first Visitation on 1 August 1713 was that large repairs were carried out on the buildings in August–September 1714.[18]

After Flamsteed died on 31 December 1719, his widow removed all the instruments from the Observatory, stating that they had all been his own property, the clocks and sextant having been given to him by Jonas Moore. The Office of Ordnance countered by threatening to sue Mrs Flamsteed to gain possession of the sextant which, they said, had been paid for by the Office.[19] However, after seeking advice from the Attorney-General, the case was dropped and, except for the clocks, none of Flamsteed's instruments have been heard of from that day to this.

EDMOND HALLEY (b. 1656, d. 1742)
Second Astronomer Royal, 1720–42

As we have seen, Halley found the Observatory devoid of instruments. The system whereby the Astronomer Royal had to find his own instruments was so obviously unsatisfactory that the Government granted Halley £500 to re-equip the Observatory.

His first action was to procure a 5-foot transit instrument which, in 1721, he fitted up in "a little boarded shed between the study and the [north-west] summer house".[20]

Flamsteed's brick meridian wall had been proved to be slowly subsiding because it was too close to the brow of the hill. About 1724, therefore, Halley built a new wall of stone—on the same meridian but standing back some yards—on which he proposed to erect two identical 8-foot mural quadrants, one looking south, the other north. Flamsteed's Quadrant House was converted into a pigeon house. Graham produced the first quadrant but the £500 proved insufficient for the second, or for the large movable quadrant Halley also asked for. To cover this meridian wall—which still survives—there was erected "a room or small house in whose roof there should be proper openings".[21] The room itself was demolished in 1749 and we have no illustration or plan of it.

JAMES BRADLEY (b. 1693, d. 1762)
Third Astronomer Royal, 1742–62

When Bradley took up residence in June 1742, he found the instruments in a sorry state, so he proceeded to rectify this as soon as possible. He says:

> "In the year 1749, 1000*l* was given by his Majesty, to be paid by the treasurer of the Navy out of money arising from the old stores of the navy, [shades of unserviceable gunpowder!] (upon the representation of the lords of the Admiralty, and principally Lord Anson's recommendation,) to buy astronomical instruments for the use of the Royal Observatory".[22]

Anson was particularly sympathetic to Bradley's needs. In 1741, off Juan Fernandez during his voyage round the world in the *Centurion,* the loss of 70 men from scurvy had been directly attributed to the fact that he had no means of finding longitude.

£500 was spent on the meridian and movable quadrants asked for in 1726 by Halley, the other half on a new transit instrument and clock, new reflecting and refracting telescopes, and on the repair of existing instruments. As well as acquiring £1000 of Admiralty funds to spend on instruments, Bradley persuaded the Board of Ordnance to build in 1749–50 the 'New Observatory' illustrated in Figs. 6 and 7, aligned to the meridian south of the present courtyard. At the west end was the Quadrant Room, built over the existing Quadrant Wall, Halley's old building being demolished. At the east end was the Transit Room, for the new transit instrument which was to define the Greenwich meridian from 1750 to 1850. In the centre were the assistant's living quarters—his bedroom upstairs, his library and calculating room downstairs. A passage connecting the three gound-floor rooms ran along the north side.

In Bradley's time or later, a new wing was built onto the south side of Flamsteed House—the first major alterations since building—surmounted by a single tall chimney which reached the height of the roof of Flamsteed House. This can be seen in Figs. 8 and 9 and is marked *11* and *12* in Fig. 7. Very few pictures or dated plans between 1700 and 1790 have survived, so many details are uncertain.

NATHANIEL BLISS (b. 1700, d. 1764)
Fourth Astronomer Royal, 1762–64

In 1762, Flamsteed's old Sextant House was converted into a small observatory with a "circular versatile roof for the reception of the 40 in. movable quadrant". This was completed after Bliss's death.

NEVIL MASKELYNE (b. 1732, d. 1811)
Fifth Astronomer Royal, 1765–1811

In 1770, Maskelyne complained to the Visitors of the inadequacies of Graham's old equatorial sector for observing comets in the Great Room. He suggested two new rooms be built, each about 12 feet square, to the east and west of the Great Room. A single new equatorial sector could, he said, be transferred from room to room according to the position of the comet.

The Visitors, in their wisdom, decided it would be better to convert the summer houses by removing the pineapple roofs, adding a storey, extending the buildings to the south, and fitting "conical movable roofs". This was done in 1773. The layout can be seen in Figs. 7 and 9.

How wrong the Visitors were! Of these new observatories, Airy had this to say in 1855:

> "The positions of these two instruments are so strangely mis-adapted to their purpose, that I am utterly at a loss to conceive under what circumstances their places were selected. The Octagon Room towers over them in such a manner that nearly the whole sky from south to east, to the height of 40° or more, is hidden from the North-west Dome; and nearly the whole sky from south to west, to the height of more than 50° in some parts, is hidden from the North-east Dome".[23]

In 1779 extensive alterations were carried out in the New Observatory: the roof shutter openings in the Transit and Quadrant Rooms were increased from 6 inches to 3 feet, the flat ceilings were removed and sloping double roofs substituted to keep the rooms cool. The 'hip' roofs seen in Fig. 6 were removed and the roof-ridges extended to the ends of the building. At the same time, the circular 'versatile roof' (or revolving dome) on the Movable Quadrant Room of 1762 was replaced by a sloping roof with sliding shutters not unlike a lean-to greenhouse. This room on the site of Flamsteed's Sextant House was henceforward known—for reasons which are not evident—as the Advanced Building and used instead of the Great Room for observations of occultations, etc. by the small telescopes.

In the Great Room the casement windows and doors were replaced by sash windows. A weather-cock (Fig. 9) was placed in the eastern roof-turret and there was a new Great Gate.

Some time after 1794, alterations were made to the living-rooms added onto the south side of the old Wren building (though exactly

8

what was done in Bradley's time and what in Maskelyne's is not clear), the dining-room, study and bedroom in Fig. 10 (known today as the Bliss and Maskelyne galleries) being built and a new entrance door outside the dining-room constructed. At the same time, according to Airy, the present courtyard was enclosed,[24] though this may have taken place in 1808 or even later.

In 1808 or 1809 the Circle Room was built onto the east end of Bradley's New Observatory. The instrument itself, replacing the two mural quadrants—respectively 87 and 62 years old—was not delivered until 1812, after Maskelyne's death.

JOHN POND (b. 1767, d. 1836)
Sixth Astronomer Royal, 1811–35

In July 1813 the new East Building was started, joining onto the eastern end of the Circle Room to provide accommodation for the increased number of assistants, and to provide a library, the old Middle Room between the Transit and Quadrant Rooms being re-designated the Computing Room.

Surmounting the East Building was a dome, originally designed for Sir George Shuckburgh's telescope, mounted as an altazimuth. The mounting proved unstable and in 1838 the dome was adapted for use with the new Sheepshanks 6·7-inch refractor.

In 1816, Bird's 8-foot transit was replaced by a new 10-foot instrument made by Troughton on the same piers.

In 1817 a large theodolite was erected on a large framework on the roof of Flamsteed House (removed 1842). In the same year a wooden Magnet House was erected in the garden for magnetic experiments. Its exact site is not certain and the following rather sad entry was made in the 1824 inventory: "The Astronomer Royal reported, that the foundation of this building had given way, and was in so dangerous a state as to make it necessary to remove the instruments. . . ."[25]

In 1818 the Admiralty took over responsibility for the Royal Observatory from the Board of Ordnance, the charge of the Royal Navy's chronometers being transferred to Greenwich in 1821. To accommodate them, the Library built in 1813 (on the first floor of the new East Building) was converted into a Chronometer Room while the Assistants' Apartments were raised another storey to form a new Library next door to the Chronometer Room—seen in Fig. 11 at top left. In 1819 a form of central heating was installed in the observing rooms and some of the living-quarters. In 1821 mains water was laid on by the Kent Water Co. The time-ball was erected on the eastern roof-turret in 1833.

GEORGE AIRY (b. 1801, d. 1892)
Seventh Astronomer Royal, 1835–81

As soon as Airy was appointed, he arranged for three new living-rooms to be added onto the west side of the ground floor of Flamsteed House—the drawing-room and two bedrooms in Fig. 10, known today as the Halley Gallery. At the same time, Bradley's single tall chimney was replaced by two of similar height close to the Octagon Room. At the same time, the Computing Room was enlarged and Quadrant Room divided (see Appendix I for the details).

Airy's régime saw a complete re-equipment of the Observatory. First, in 1836–7, a new wooden magnetic observatory was built in a newly enclosed area to the south of the main buildings, just south of today's Altazimuth Pavilion. Aligned to the magnetic meridian and constructed entirely of non-magnetic materials, it was of cruciform shape and can be seen in Fig. 13.

Next, in 1844, he built the three-storey Altazimuth Dome on the walls of Flamsteed's old observatory and the later Advanced Building. The low wooden drum-dome can be seen from the south and north respectively in Figs. 48 and 12.

Then, in 1849, work started to convert the Circle Room to accommodate the new transit circle, replacing Pond's transit instrument and mural circle. This was completed in 1851—and the Greenwich meridian was thereby shifted 19 feet eastward, to be recognized as the world's Prime Meridian in 1884. Bradley's Transit Room became the Astronomer Royal's official room.

His last major work was the building of the new south-east dome east of the East Building for the 12·8-inch Great Equatorial telescope. The three-storey octagonal building was built in 1857, the telescope erected in 1859. The drum-dome can be seen behind the hemispherical Sheepshank dome in Fig. 12 and on the right of Fig. 83.

Other building alterations in his time included a covered way from the house to the Observatory in 1840; the erection of an external staircase on Flamsteed House in 1849 to give direct access from the courtyard to the roof where meteorological equipment was mounted; the installation of gas lighting in 1851; the building of the two fireproof record rooms onto the east end of the East Building in 1854–5; the erection of a wooden dome on the south ground for a photoheliograph in 1873; a new Library near the Magnetic Observatory, completed in 1881 (right in Fig. 13).

WILLIAM CHRISTIE (b. 1845, d. 1922)
Eighth Astronomer Royal, 1881–1910

The advent of two new techniques in the nineteenth century, photography and spectroscopy, made possible a new branch of astronomy—astrophysics—concerned not so much with the motions as with the composition of the heavenly bodies. Airy had said in 1875: "the Observatory is not the place for new physical investigations".[26] His successor, however, immediately set about making plans for bringing Greenwich into the forefront of astrophysical research, resulting in a great enlargement in the scope of the work.

Christie's first achievement was to equip the Royal Observatory with a comparatively large telescope. In 1885 the decision was taken to replace the Merz 12·8-inch Great Equatorial refractor with a 28-inch telescope on the same mounting. This necessitated a new onion-shaped dome (Figs. 13 and 88), completed in 1893.

As Greenwich's contribution to the international *Carte du Ciel* project, a 13-inch astrographic telescope was mounted in 1890 in a new dome above Halley's old quadrant pier at the west end of Bradley's observatory (Fig. 15 centre).

The New Physical Observatory—known today as the South Building—was started on newly enclosed ground south of Airy's Magnetic Observatory in 1891. The three-storey terracotta building, completed in 1899 (Fig. 16), was cruciform in shape with four wings and a central tower carrying a 30-foot dome for the Thompson equatorial (originally erected at ground level for the Lassell reflector). The architect was Frank Crisp. This new building provided greatly needed office, computing and workshop accommodation.

Turning to fundamental astronomy, Christie acquired a new altazimuth instrument, a new pavilion in the same terracotta style being built just north of the old Magnetic Observatory in 1899. Soon after, a new enclosure—the Christie enclosure—was made in the park some 350 yards east of the old observatory where a new magnetic pavilion was built, away from the disturbing effects of iron and steel in the new buildings.

11

Minor building projects in Christie's time included a porter's lodge with a high chimney in 1889, the Transit Pavilion on Bradley's meridian in the front courtyard in 1891, electric lighting from 1893, telephones from 1886, a new covered way between Flamsteed House and Bradley's observatory in 1908.

FRANK DYSON (b. 1868, d. 1939)
Ninth Astronomer Royal, 1910–33

Soon after Dyson's appointment, the accommodation in Flamsteed House was enlarged for the last time by the construction in 1911 of new basement rooms at the west end of the house—known today as the Spencer Jones Gallery—immediately under the rooms added by Airy in 1835.

In 1911, Airy's altazimuth was dismounted from its dome on the site of Flamsteed's observatory and replaced by a Dallmeyer photoheliograph. The drum-dome opening was slightly modified. In the same year a wooden hut was erected in the courtyard for the Cookson Floating Zenith Telescope.

Airy's wooden magnetic observatory on the South Ground was demolished in 1917.

In 1923 the electrification of the Southern Railway suburban system forced the Observatory to transfer its magnetic observations to a new site at Abinger in Surrey.

In 1931 the former magnetic observatory in the Christie enclosure was demolished and two new buildings erected there: a 34-foot dome for the Yapp 36-inch reflector, and a pavilion for the Cooke Reversible Transit Circle. Both instruments came into use after Dyson's retirement.

HAROLD SPENCER JONES (b. 1890, d. 1960)
Tenth Astronomer Royal, 1933–55
RICHARD V. D. R. WOOLLEY (b. 1906)
Eleventh Astronomer Royal, 1956–71

In 1939, deteriorating astronomical conditions—bright street lights, fog, deposition of dust particles on mirrors and pivots—forced the Astronomer Royal to recommend that a new site should be found for the Observatory. However, before any action could be taken, war broke out. The objectives of the larger telescopes were dismounted and sent to a place of safety; some departments virtually closed down for the period of the war, never to return to Greenwich;

some continued to function elsewhere; only the Solar, Astrometric and Meteorological Departments remained at Greenwich. Of the telescopes, only the Dallmeyer photoheliograph and Sheepshanks refractor continued to function, though the latter was rarely used.

In October 1940, bombs destroyed the main gates and caused considerable damage to the coverings of the domes, as well as damage to the Altazimuth Pavilion and the small transit mounted in it. There was also considerable blast damage to the buildings. Fig. 17 was taken soon after this.

In July 1944 a V1 flying-bomb caused widespread superficial damage, stripping most of the covering off the 28-inch dome.

During the war, considerable research had been carried out to find the best possible future site for the Observatory. In 1945 the Board of Visitors unanimously recommended that Herstmonceux in Sussex should be chosen and the Admiralty announced this in April the following year, saying that the conditions for astronomical observations there appeared to be as good as could be obtained in England. Spencer Jones, in reporting this, continued: "Herstmonceux Castle, built in 1446 by Sir Roger de Fiennes, Treasurer to the Household of Henry the Sixth, is probably the most beautiful early brick building in the country and will provide a dignified future home for the Royal Observatory, appropriate to its long history and tradition".[27]

In the meantime, the war in Europe had come to an end and, in July 1945, the Astronomer Royal and his administrative staff, who had been at Abinger since the autumn of 1940, returned to Greenwich, the Astronomer Royal and his family once more taking up residence in Flamsteed House.

In 1948 the move to Herstmonceux began, a move which, to the frustration of all concerned, was to take almost ten years. In August the Astronomer Royal and the Secretariat moved from Greenwich and in September the Chronometer Department moved from Bradford-on-Avon. In the same year the name of the Observatory was changed by Royal Assent from the Royal Observatory, Greenwich, to the Royal Greenwich Observatory, Herstmonceux.

As the observatory staff moved out, so the care of the buildings at Greenwich was taken over by the Ministry of Works (now the Department of the Environment). In 1951 the final decision was made that the old buildings in Greenwich Park should be progressively handed over to the National Maritime Museum "to be used as an astronomical and navigational annexe"[28]—a most appropriate decision in view of the Observatory's nautical beginnings. At the same time, the Astronomer Royal transferred to the museum's care many of the historic instruments described in this book.

As we have indicated, the move proceeded all too slowly. The 28-inch and Thompson 26-inch/30-inch equatorial were dismounted and sent to Herstmonceux in 1947, the Photoheliograph in 1949, the Reversible Transit Circle in 1953, the Yapp reflector in 1955 and the Astrographic refractor in 1956. The old Airy Transit Circle, defining the prime meridian of the world, was left in working order in the museum's care, the last regular observations having been taken in 1954. The Sheepshanks refractor was also left in working order, being moved from its old dome in the East Building to Christie's Altazimuth Pavilion in 1963.

The Solar Department left Greenwich and the *Nautical Almanac* left Bath in 1949. The Time Department finally left Abinger in 1957. Other departments were split between Greenwich, Abinger and Herstmonceux for several years. What seems to have been the final event in this long-drawn-out process took place in October 1957—the moving of some 30,000 glass photographic plates taken with the 26-inch and astrographic telescopes over the previous 60 years, not one of which was broken. Except for the descriptions of individual instruments, the development of the Royal Greenwich Observatory and its buildings at Herstmonceux will not be further discussed here, being referred to elsewhere.[29] Meanwhile, plans were being made for the converting of the historic buildings at Greenwich into a museum—first by demolition. During the 1950s, the RTC, Yapp and Cookson domes in the 1900 Christie Enclosure in the park were demolished to such effect that no trace now remains. In the main buildings, the post-war gate-house (built after the original had been demolished by a bomb), transit pavilion, new library, astrographic and photoheliograph domes, and the framework of the old 'onion' dome of the 28-inch refractor were all swept away.

On the positive side, the Octagon Room—in which were exhibited Halley's transit and Bradley's zenith sector—was opened to the public by H.R.H. Prince Philip in May 1953, the Airy transit circle also being made accessible to the public. In 1955 the whole of Flamsteed House—residence of successive Astronomers Royal for 272 years—was taken in hand for restoration by the Ministry of

Works, being opened as part of the National Maritime Museum by H.M. The Queen on 6 July 1960.

On 17 November 1965, H.R.H. Prince Philip (one of the museum's trustees) opened the Greenwich Planetarium, in the old Thompson Dome surmounting the South Building. Seating about 50, regular programmes are given to schools and the general public throughout the year. In 1965 also, the Ministry of Public Building and Works took in hand for restoration Flamsteed's observatory at the bottom of the garden and today's Meridian Building—Bradley's 'new observatory' of 1750, the transit circle room and East Building of about 1810, and the Record Room of 1870 (Fig. 18).

It was decided that, as far as possible, each part should be restored to its state when in active astronomical use and that the ancient instruments should be re-instated in their working positions. This involved not a little historical research to discover just what those states had been, the research being carried out jointly by the museum, the Observatory and the Ministry, the fruits of which can be seen not only in the buildings themselves but also in the present volume.

The Meridian Building was opened to the public by Sir Richard Woolley, Astronomer Royal, in July 1967.

In November 1971, the 28-inch refractor of 1893 with its mounting of 1859—at Herstmonceux since 1947 but having now come to the end of its useful professional astronomical life—was re-erected in its original place on the top of the Great Equatorial Building at Greenwich. At the time of writing (1974), a new working 'onion' dome—of the same appearance as that damaged in the war but of modern materials—is being erected over the telescope. By 1975 this famous telescope—still the seventh largest refractor in the world— will once more be in full working order, the largest telescope available to the public in the United Kingdom.

Britain's oldest scientific institution was founded by King Charles in 1675 "for perfecting navigation and astronomy". In the tercentenary year of 1975, this function is still being performed at Herstmonceux by the Royal Greenwich Observatory. At Greenwich the historic buildings have been restored as part of the National Maritime Museum so that the visitor can see seventeenth, eighteenth and nineteenth-century observatory rooms almost in their original state, and can also, for good measure, stand astride Longitude Zero and hear the Greenwich time signal—still based on the Greenwich Meridian but now emanating from Herstmonceux.

NOTES

1. Baily, pp. 37–8.
2. Baily, p. 126.
3. Translation of *H.C.*, III (1725), p. 101.
4. Baily, p. 39.
5. RS CM. 1/239.
6. Baily, p. 39n; Airy, *Astron. Obs.* 1862, Appendix II, p. 9.
7. Baily, p. 190.
8. RS MS. Vol. 243 (F1), Lr. 12.
9. *Wren Society*, V, 21, in a letter of 1681 to Bishop Fell concerning the building of Tom Tower at Oxford.
10. RS MS. 243 (Fl), Lr. 12.
11. PRO/WO 55/391/120, 16 September 1675.
12. PRO/WO 47/19 B., 9 November 1675.
13. J. R. Tanner, *Descriptive Catalogue of naval manuscripts* . . . IV (1923), pp. 280, 293.
14. PRO/WO., 47/19 B/16, September 1675.
15. RGO MS. 40/62^r.
16. RGO MS. 2/168–71.
17. RGO MS. 4/58.
18. RS MS. Flamsteed to Sharp, Lr. 90.
19. PRO/WO 47/33/207–8.
20. Baily, p. 343.
21. *ibid.*, p. 21.
22. Rigaud, lxxv.
23. G. B. Airy, *Address* . . . (1855), p. 3.
24. Airy, *Astron. Obs.*, 1862, Appendix II, p. 15. A water-colour dated 1794 in possession of N. Arnold Forster, Esq. shows that the alteration must have taken place after this.
25. RS MS. 371/28/7.
26. Airy, *Report*, 1875.
27. Spencer Jones, *Report*, 1946/25.
28. Spencer Jones, *Report*, 1952/28–9.
29. Science Research Council, *The Royal Greenwich Observatory Herstmonceux* (1969).
   W. H. McCrea, *The Royal Greenwich Observatory: an historical review issued on the occasion of the tercentary* (1975).
   A. J. Meadows, *Greenwich Observatory 1835–1975* (Taylor & Francis, 1975).

2

The Mural Instruments

THIS chapter describes those Greenwich instruments, successors to Tycho Brahe's *Quadrans Muralis*, consisting of a graduated arc (which may be a quadrant, a full circle, or some arc in between) fitted with a sighting device—either open sights (as Tycho's) or a telescope (as Flamsteed's)—mounted on a wall or pier aligned exactly in the meridian. With such an instrument, the astronomer could measure the zenith distance (ZD) of a heavenly body as it crossed the meridian, the sidereal time of transit yielding the right ascension (RA). Knowing the precise latitude, he could derive declination (Dec), often expressed as North Polar Distance (NPD).

All the instruments described were originally designed to yield both RA and Dec from one observation. However, the increasing need for precision forced the astronomers to use the later mural instruments to measure ZD only, a separate transit instrument being used to provide RA. These mural instruments were superseded by the transit circles with which both RA and Dec could be measured simultaneously.

2.1. *Hooke's 10-foot Mural Quadrant* (1676)

Designed by Robert Hooke; made by Thomson of London; divided by John Flamsteed. Mounted 1676; no record after 1678.

Description (see Fig. 19)

Dividing the whole ninety degrees of the limb of a quadrant of large radius is a laborious business, open to many inaccuracies. The ingenious Robert Hooke (1635-1703) tried to simplify matters by

17

marking the limb itself every 5° with brass studs and having a fully graduated subsidiary arc 5° long, *D* (Fig. 19), which could be clamped to the limb with its zero on whatever 5° mark was appropriate to the altitude of the body to be observed, rather as the vernier scale is clamped to the limb of a marine sextant.

To make an observation, the subsidiary arc *D* was first clamped to the limb as described above. The telescopic sight *BAC* was then moved by the crank *E* to bring the object onto the horizontal wire of the telescope (inside *A*). The reading of the index *A* on the scale *D*, added to that of index *D* (an even 5°), gave the required ZD measurement.

In practice, great difficulties were encountered in forcing the quadrant into the meridian. More serious, the double index proved impossibly cumbersome, as witness Flamsteed's complaint to Sir Jonas Moore in July 1678: "... I tore my hands by it and had like to have deprived Cuthbert [Denton] of his fingers". Well might he add: "... except something more manageable may be put in its place, it will be a great let to our proceedings. ..."[1] The radius of the quadrant was 10 feet. The object-glass (aperture not known) cost 3*s*. 0*d*, the cell for it 12*s*.; the eyeglass 4*s*. 6*d*.[2]

Historical Summary

The sextant having cost more than expected, Flamsteed's own design for a mural arc or semicircle was rejected by Sir Jonas in favour of Hooke's double-index quadrant,[3] made in London by Thomson, being complete by 6 May 1676.[4] By July 1676 it had been mounted, under Hooke's supervision, on the east wall of the Quadrant House at Greenwich. It was then handed over to Flamsteed who spent much time and effort on it before abandoning it as useless. It disappears from the record after December 1678.

The following, written by Flamsteed in 1682, gives some idea of the relationship between Moore, Hooke and Flamsteed:

"The common way of observing the Sun's place has been by his meridional heights observed by a good large quadrant. Such a one I moved to have had made; but Sir Jonas would needs have Mr Hooke contrive it; which he did without any consent of mine, but so ill that it was impossible for me to render it useful, though I employed my utmost endeavours to make it serve: which causes me to think he ordered it so on purpose to hinder my progress."[5]

2.2. *Flamsteed's 'Slight' Mural Arc* (1681)

Made in 1681 to Flamsteed's design by local workmen and mounted 1683. Abandoned 1686. Parts probably used for next instrument.

Description

Arc more than 140° so that all stars visible at Greenwich could be observed. Radius 6 feet 9 inches. Probably generally similar to next instrument.

Historical Summary

Having failed to make Hooke's quadrant work, Flamsteed started to make a mural arc at his own expense in August 1681. It was finished the same year but, through bad workmanship, the limb was faulty[6] and it was not mounted until 1683. Hooke's quadrant had been on the east wall of the Quadrant House: this was on the west wall. Flamsteed used it for taking meridional heights from 1683 until autumn 1686 when he abandoned it because, he said, "it was built too slight, and could not be well fixed: so I durst not confide in the measures taken with it . . .".[7]

2.3. *Flamsteed's Mural Arc* (1689)

Made in 1688 by Abraham Sharp. Mounted on west wall of Quadrant House by July 1689. Removed 1720. No record after 1721.

Description (key to Fig. 20)

IKRr: Brass limb, 140° long, 6-foot 7½-inch radius to screwed outer edge. Engraved scale reading ZD to 5″ of arc on diagonal scale, or to 1/100 of a revolution of the endless screw with the aid of a micrometer head.[8] A table for converting 'revolves' into degrees, minutes and seconds was published.[9] The zero of the scale was checked periodically by plumb-line.

C to eye: Telescopic sight, 7 feet long, with convex lenses and crossed threads, fixed to a brass bar, the whole pivoting at C.

Index, fixed to eye-end of telescope, moved round the arc by a crank working an endless tangent screw, which could be disconnected when fast movement was needed.

CL, CI: Heavy iron radii.
Cr, Cr, Cr: Lighter iron radii.
EK, rL: Heavy iron bars.
db, db: Strengthening bars.

N, N, E: Oak blocks fixed to the wall, on which the frame is suspended. It was kept in the meridian by the use of wedges working on wooden blocks under the circumference.

Counterpoise apparatus (not shown) for index and telescope, consisted of rope, pulleys and balance weight.

Roof opening. A slit 1½ feet wide, allowing stars down to magnitude 7 to be seen with the naked eye.[10]

Method of use

1. Set index of telescope to approximate ZD of star to be observed. The endless screw can be disconnected to allow this to be done.
2. Look through telescope and wait for star to come into field of view from the right.
3. Bring star onto horizontal cross-wire by moving index crank-handle.
4. Note time of transit (when the star reaches the vertical crosswire) to nearest half-second by counting the beats of the clock.
5. Record clock-time of transit in column (A) (Fig. 21).
6. Read off observed zenith distance in degrees, minutes and seconds from the diagonal scale and record in column (D), and in revolutions and hundredths of the endless screw and record in column (E).

Completing the Record (Fig. 21)

7. Record true apparent time in column (B) by applying known clock-error to clock-time in column (A). This is only necessary for Sun, Moon and planets.
8. Convert revolutions and hundredths in column (E) into degrees, minutes and seconds of arc, using a conversion table.
9. Obtain corrected zenith distance to enter in column (F) by applying the known zero error (G) to the value in column (D) and (E) combined.

Historical Summary

As we have seen, Flamsteed had no great success with his first two mural instruments. His presentation to the living of Burstow (Surrey) in 1684 and his father's death in 1688, however, improved his financial position and he was able to commission a new mural arc, apparently incorporating certain parts of his previous instrument.

Starting in August 1688, most of the work was done by Abraham Sharp (1653–1742) his assistant, who

> "strengthened the edge of the limb with screws, engraved the degrees, fitted the index, and contrived all and each part of it so ingeniously that this piece of work was admired by all the expert craftsmen who saw it".[11]

All of which cost Flamsteed more than £120 in cash and 14 months in time. Finished in October 1689, mounted on the west wall of the Quadrant House, Flamsteed at last had the fundamental instrument he had been wanting since 1676. With it he obtained most of the data needed for his great catalogue of 2935 stars published after his death in Volume III of *Historia Coelestis Britannica*.

Between 1689 and 1719, some 28,650 observations are recorded, the last being on 27 December 1719, four days before Flamsteed's death at the age of 74. His widow removed the instrument in 1720. On 20 April 1721, Joseph Crosthwait, Flamsteed's last assistant, wrote to Sharp, "Mr Molyneux has a mind to purchase the mural arc, and the [movable] quadrant that used to stand in the great room".[12]

Alas, this instrument—so famous in the history of British astronomy—has not been heard of since. Luckily, however, a fine contemporary picture has survived not a mile away from the place where Flamsteed and his staff used it—on the ceiling of the Painted Hall of today's Royal Naval College, at the bottom of the hill not far from the river. Painted from life by Sir James Thornhill in 1714, (Fig. 22), it shows an excellent portrait of Flamsteed with his then assistant, Thomas Weston, standing in front of the mural arc.

Contemporary Accounts
1718: RGOMS 32/67. 1718: J. Flamsteed, *Historia Coelestis Britannica*, III Prolegomena, 108–11. 1710: Z. C. v. Uffenbach, *Mertwürdige Reisen Durch Wiedersachsen, Holland und Engelland*, II (Frankfurt & Leipzig, 1753), 444–5, translated into English by W. H. Quarrell and M. Mare, *London in 1710* (1934).

2.4.   *Halley's 8-foot Iron Mural Quadrant (1725)*
The prototype for similar instruments used all over Europe
Constructed under the direction of George Graham.
Mounted on new stone pier (facing south), 1725.

Re-divided by John Bird, 1753 and re-mounted facing north.
Dismounted, 1888.
Displayed in Quadrant Room, Old Royal Observatory, since
1967.

Description

This quadrant, designed by George Graham (1673–1757), the
famous clockmaker, served as a prototype for similar instruments
used in observatories all over Europe after 1750. Indeed, the
flourishing export market achieved by English instrument-makers
from that date owes much to this instrument of Graham's. The two
main factors leading to its success were great rigidity and great
accuracy, Graham dividing the 8-foot radius limb with a degree of
accuracy unknown before,[13] giving it two sets of divisions, readings
from each of which could be checked one against the other (as
Flamsteed had checked his 'revolves' against the degree scale of the
sextant and mural arc).

The inner scale on the limb was divided conventionally into 90°,
subdivided by vernier—Flamsteed had used a diagonal scale—to
30″ and, from 1745 by micrometer to 1″. The outer scale was divided
into 96 parts, subdivided by vernier into 1/256 parts, and by
micrometer to 1″.

The division of the quadrant into 96 parts was achieved by con-
tinual bisection of the 60° arc, which was laid off by inscribing an
arc equal to the radius of the quadrant and which was numbered 64.
Successive bisections then gave 32, 16, 8, 4, 2, 1, etc. parts. One part
was thus equal to 56″·15. A table was produced for converting the
parts of the outer scale into degrees, minutes and seconds. The outer
scale reading could thus be compared directly with the inner scale
reading.

The main parts of the instrument as described here refer to Fig. 23
which is actually a picture of a later export model, from a French
encyclopedia. The dimensions and other details quoted refer,
however, to Graham's own instrument at Greenwich. Notes in
parentheses indicate the state of the instrument as mounted in the
old Royal Observatory today.

Key to Fig. 23
AC: Plumb-line. Wooden plumb-line guard and plummet pot
were added in 1786 (all missing).

BC: Brass limb, radii 7 feet 11·75 inches to 90° arc, 8 feet 0·77 inches to Graham's 96 arc, 8 feet 0·22 inches to new 96 arc by Bird.[15]

FH: Telescope 8-foot 0·5-inch focus. Before 1789, common OG with aperture restricted to 1·4 inch, magnification ×51. After 1789, achromatic OG. Collimation checked with zenith sector.

a, b: Suspension points for frame.

abcd: Iron counterpoise crank (missing).

efg: Mahogany counterpoise frame, attached at the top to the crank *cd*, at the bottom to the eye-end of the telescope at *e* (frame missing). Metal braces to prevent flexure of telescope (omitted from drawing in Fig. 23) but seen in Fig. 25.

hi: Iron counterpoise weight attached to crank at *b*.

i: Point on scale against which plumb-line must lie.

k: Brass vernier index attached to eye-end of telescope, moved over scales on limb by tangent screw *op*.

mn: Brass clamp (modern replacement).

op: Tangent screw, fitted with micrometer head in 1745.

Method of use
    See p. 25.

Historical Summary

1724: Halley received grant of £500 to re-equip Observatory; quadrant ordered from Graham; meridian wall of nine stones only erected just north of Flamsteed's brick meridian wall. 1725: quadrant made by Jonathan Sisson under Graham's direction. July: dividing carried out in Great Room by Graham himself.[16] September: erected in new Quadrant Room, on east side of new wall; cost reported as follows:

The mason's work for stoneworks of the walls	£42	17	0
The bills for making the Quadrant in which are contained several particulars common to that & the other Quadrant, not put together and such part of that other Quadrant itself as are provided	£204	8	6
The Bricklayers, Carpenters, Plumbers, Glaziers, Painters and Smiths' bills, for building & fitting up the house that contains the said Quadrants	72	1	10
A plain Month Clock to stand by the Instruments	12	0	0
The glasses for the Telescope	3	3	0[17]

The second quadrant mentioned was not completed, the money granted having been expended.

27 October 1725: first recorded observation; Halley used his quadrant for measuring both RA and Dec, more or less abandoning his transit instrument from this time. 1742: on Halley's death, Bradley found the counterpoise was rubbing on the ceiling.[18] 1745: micrometer fitted to telescope by John Bird; 3-inch focus eyepiece replaced by one of 2-inch focus. 1749: Quadrant Room demolished but Quadrant Wall left standing with Halley's Iron quadrant still mounted; Bradley's New Observatory built around Quadrant Wall. 1750: new quadrant mounted on west side of pier, of exactly the same design except that frame is of brass instead of iron. 1753: iron quadrant dismounted from east side of pier; limb re-divided by John Bird; re-mounted on west side of pier, facing north (as at present). 1775: gold points let into centre plate and limb for adjusting plumb-line. 1780: movable eyepiece applied to telescope. 1786: wooden guards applied to plumb-line "to defend them from the agitation of the air": an extra 5 lb applied to counterpoise. 1789: achromatic object-glass fitted by Peter Dollond.

1 August 1812: last recorded observation; Mural Circle brought into use. 1839: Quadrant Room converted into Muniment Room; roof openings and iron quadrant remained in place after being "put in such order [by Simms] that, if any future astronomical antiquary should desire, observations may be made with it".[19] 1888: dismounted from Quadrant Wall and hung on west wall of Transit Circle Room as a relic (see Fig. 73).

1952: dismounted from Transit Circle Room, restored in museum workshops. 1967: mounted in 1753 position.

Contemporary Accounts
1738: R. Smith, *A Compleat System of Opticks* (1738), Bk.4, 332–41. *c.* 1772: RGOMS 251/36r. 1776: N. Maskelyne, *Astron. Obs.*, I (1776), i.

2.5. *Bradley's 8-foot Brass Mural Quadrant* (1750)
By John Bird to Graham's design.
Inscribed on limb: *John Bird London.*
Mounted Quadrant Room, 1750–1952.
Displayed in Old Royal Observatory, Quadrant Room from 1967.

Description (Figs. 25 and 26)

Identical to Graham's Iron Quadrant except that the frame is made of brass. A contemporary picture of it can be seen in Fig. 75.

Method of use (a planet) (Fig. 24)

1. Clamp telescope index at expected ZD.

2. Look through telescope and wait until planet enters field of view from the right.

3. Bring image of planet onto horizontal cross-wire by turning micrometer screw, then:

4. Read observed ZD at index on both exterior and interior divisions and record them as in columns (A) and (C) of Fig. 24.

5. Convert exterior division reading (A) into degrees, minutes and seconds (B), using conversion table.

Historical Summary

As we have seen, the money in Halley's £500 grant of 1724 ran out before the second—north-facing—quadrant could be completed. His successor, James Bradley, was granted £1000 for further re-equipment in 1748 and gave an order for a second quadrant to John Bird in February 1749.

Graham's quadrant had been acclaimed by all, the only fault to be found with it was that it had altered its figure by its own weight so that the arc was now 16″ less than 90°. Bird overcame this fault in the new quadrant by making the frame of brass instead of iron. Otherwise, the two quadrants were virtually identical—and interchangeable.

Bird described the quadrant itself and the method of dividing it in the two pamphlets mentioned below, published by the Board of Longitude. These pamphlets proved to be important in the subsequent selling of similar quadrants elsewhere, especially abroad. Bird received £500 from the Board as a reward for writing the pamphlets and making known his methods.

The subsequent history of the brass quadrant is summarized as follows: 1750: mounted on west side of Quadrant Wall, facing north; first observation 10 August. 1750: counterpoise fitted; telescope balanced. 1753: iron and brass quadrants changed places; brass now on east side of wall, facing south. 1769: Jean Bernoulli reported a diagonal eyepiece being used.[12] 1772: achromatic object glass and removable eyepiece fitted by Peter Dollond; new level by Nairne. 1775: plumb-line guard fitted to "defend it from the agitation of the air"; gold points applied to limb for use with new gilt-silver plumb-line. 1779: roof openings widened. 1787: new micrometer screw "according to Mr. Smeaton's method, which being the most exact, I intend to use in future".[22] 1789: micrometer modified by Troughton.

11 August 1813: last regular observation; mural circle brought into use. 1824: used while Troughton Mural Circle was boxed up during erection of Jones Mural Circle. 2 January 1825: last recorded observation. 1839: Quadrant Room divided; Brass Quadrant remained on east side of wall. 1890: wall used as part of support for Astrographic Telescope.

1952: quadrant dismounted and restored in Museum workshops. 1967: quadrant mounted in 1753 position. Displayed in ORO.

Contemporary Accounts
   1768: J. Bird, *Method of Constructing Mural Quadrants* (*1768*).
1767: J. Bird, *Method of dividing Astronomical Instruments* (*1767*).
1700: N. Maskelyne, *Astron. Obs.* (1776), I, vi–ix.

2.6.   *Troughton 6-foot Mural Circle* (1810)
By Edward Troughton of London. The optical parts by P. & J. Dollond of London.
Signed on limb: *Troughton London 1810.*
Mounted in Circle Room, 1812–50.
Now in Pond Gallery, Old Royal Observatory.

Description (Figs. 28–30)
   The mural circle differed from previous mural instruments in three main particulars:

(*a*)   in having a scale of 360° instead of 140° or 90°;

(*b*)   in having a moving scale with six fixed indices instead of a fixed scale with a single moving index;

(*c*)   in dispensing with the plumb-line for defining the zero; initially, Pond measured NPD direct by treating the circle as a differential instrument and using the circumpolar stars to define the pole; after 1825 when he had two circles, a mercury surface provided the horizontal for one of the two simultaneous observations of each body: (the zenith micrometer on the back of the pier, seen in Fig. 28, was intended to provide a plumb-line reference, but it was defective and never used).

The form of the instrument can be seen in Fig. 28. The telescope is clamped to the 6-foot circle which can be rotated on its conical axis, set into an aperture of a 4-foot thick wall (now the east pier of Airy's transit circle). The back bearing of the axis had two adjustments at right-angles to make the circle precisely vertical and precisely in the meridian.

With divisions every 5' of arc, the scale is engraved on a white-metal band containing four parts of gold and one part of palladium let into the rim of the circle. The figures which count the degrees are engraved on a ring of platinum. These materials were chosen because neither will tarnish, thus obviating the cleaning which in time wears out the fine divisions. Micrometer 'A' reads NPD direct (increasing from 0° above the Pole, decreasing from 360° below the Pole).

Fixed to the pier at 60° intervals are six reading-microscopes pointing inwards towards the scale on the rim of the circle. The 5' divisions of the scale can be subdivided to single seconds by the wires of the microscopes, and estimated to a tenth of a second by the micrometer head.

As we have seen, the telescope and the circle rotate together when in use. As originally designed, it was expected that the position of the telescope on the circle would be varied at intervals to permit the calculation of instrumental errors. However, these errors proved to be so insignificant that after the first two years the telescope remained clamped at 0°.

In 1825 a second circle—by Thomas Jones and almost identical with Troughton's—was mounted in the Circle Room facing Troughton's circle so that simultaneous observations of the same body, direct and by reflection, could be taken simultaneously by the two circles, one observing the direct image, the other the image reflected from a mercury trough placed on a shelf under the circle as in Fig. 29.

Only the moving parts of the instrument have survived—the circle with the telescope clamped to it and the inner conical axis. The parts originally attached to the pier have all disappeared—notably the six reading microscopes, the three clamps with micrometer adjustment used for setting the circle, the four mahogany arms to place the lamps on, and the counterpoise apparatus for lifting the front of the circle almost off its bearings.

The reading-microscope is copied from the Edinburgh Mural Circle, preserved in the Royal Scottish Museum and virtually complete.

The original circle pier is now the east pier of Airy's transit circle.

Historical Summary

With the development of the dividing-engine towards the end of the eighteenth century, the full circle (as opposed to, say, a quadrant) became popular, particularly for theodolite-type instruments such as Piazzi's 5-foot Palermo circle by Jesse Ramsden, completed in 1789.

Maskelyne asked for a meridian circle in 1792 but no action was taken until 1806 when he raised the matter once more:

> "The Astronomer Royal represented to the Council of the Royal Society that, in consequence of the great improvements that have been made in the last twenty years, principally by British artists, in the construction of astronomical instruments, most of the Observatories in foreign countries are now furnished with divided circles for observing the distances of celestial objects from the Zenith; and that from the nature and construction of such instruments, observations made with them may be expected to be more accurate than those made with a mural quadrant, however large its dimension may be."[23]

This renewal of interest occurred largely as a result of observations by John Pond, future Astronomer Royal, at his home in Westbury, Somerset. With a 30-inch altazimuth by Edward Troughton, he concluded that Bird's quadrant at Greenwich could no longer be trusted.

The subsequent history may be summarized as follows: 1806: Troughton showed a model of a mural circle designed to replace both quadrants; 6-foot radius decided upon. 1807: mural circle ordered from Troughton at request of the Astronomer Royal, Nevil Maskelyne; he suggested that it might possibly replace the transit instrument (also 55 years old) as well. 1808–9: Circle Room built immediately east of Transit Room. (In 1850 this was converted into the Transit Circle Room.) 1811: death of Maskelyne; Pond became Astronomer Royal. 1812: mural circle erected; the circle cost £735; the optical parts £53 6s.; the mason's work £28 8s. 6d. 11 June 1812: first observation: 1813: mural circle did not prove steady enough to be used for transits, so a new transit instrument was ordered from Troughton. 1821: method of clamping telescope to circle changed after errors had been detected. 1822: start of series of observations of stars by reflection in a bath of mercury; this method was so successful that a second circle, which had been destined for the Cape of Good Hope, was installed in 1824, facing the existing one. 27 January 1825: start of observations by both circles. 1839: Jones Circle sent to Cape; observations continued with Troughton Circle alone; the new Astronomer Royal, Airy, found it possible to take observations direct and by reflection both on the same transit. 1848: mounted and used on east end of East Building while Circle Room was converted into Transit Circle Room. 29 December 1850: last observation, superseded by Airy's Transit Circle. 1851: hung on east wall of Transit Circle Room as relic.

1967: mounted in present position, about 13 feet east of its old working position.

Contemporary Account

1829: W. Pearson, *An Introduction to Practical Astronomy*, II (1829), 472–88. 1837: *Penny Cyclopaedia*, VII (1837), 188–9, Article 'Circle'.

2.7. *Jones 6-foot Mural Circles* (1821 and 1822)

Both by Thomas Jones of London. Object-glasses probably by Charles Tulley.[24] Present whereabouts unknown.

As we have seen, Pond was so enthusiastic about the idea of using two circles simultaneously that he was able to persuade the Admiralty in 1823 that the circle—similar in all respects to the one at Greenwich—being made by Jones for the newly founded Royal Observatory at the Cape of Good Hope (which we will call Jones I) should be erected at Greenwich, a third circle (Jones II) being ordered for the Cape.

Both Jones circles seem to have been virtually identical with Troughton's circle at Greenwich, except that the scale of Jones I was of gold instead of gold/palladium.[25]

Jones I, which cost £411 3s. 2½d., was erected at Greenwich in 1824, the first recorded observation being on 27 January 1825. It remained in constant use at Greenwich until 1839 when it was sent to the Cape because of complaints of faulty divisions and ill-shaped pivots on Jones II.

Jones II was erected at the Cape Observatory about 1828 and shipped back to England in 1839 for the reasons given above, being then mounted on Jones I's pier at Greenwich.

It was used at Greenwich only for two months in 1848 while the Troughton circle was being moved to a temporary position while the Circle Room was being converted to accommodate Airy's new transit circle.

In November 1851, Jones I was sent to Queen's College, Belfast. Its present whereabouts is not known.

NOTES

1. RGO MS. 36/62, Baily, p. 118; Flamsteed to Moore, 16 July 1678.
2. RGO MS. 2/168.
3. RGO MS. 35 (Baily, p. 43).
4. Hooke diary, 6 May 1676 (Robinson, 23).
5. RGO MS. 42, Baily, pp. 127–8; Flamsteed to Sherburne 12 July 1682.
6. RGO MS. 33 (Baily, p. 51).
7. RGO MS. 41 (Baily, p. 54).
8. RGO MS. 32/38 (written 1718).
9. *HC* II, [2] to [15].
10. Baily, p. 57.
11. *HC* III, Prolegomena, p. 108.
12. RS MS. Baily, p. 343, Crosthwait to Sharp, 21 April 1721.
13. N. Maskelyne, *Astron. Obs.*, 1766–1774 (1776), I, 1.
14. *Ibid.*, Table LVI, p. [48].
15. RGO MS. 251/3.
16. RS MS. Baily, p. 359; Crosthwait to Sharp, 24 July 1725.
17. ROV Vol. I, pp. 23–4.
18. Rigaud, p. 381.
19. G. B. Airy, *Report*, 1839.
20. ROV Vol. I, p. 32.
21. Bernoulli, 81.
22. Maskelyne, *Astron. Obs.*, III, Quadrant, 3, 15 March 1787.
23. ROV Vol. II, 11 July 1806.
24. Pond, *Astron. Obs.*, 30 January–March 1825.
25. *The Mirror* . . . 30 January 1830, p. 83. Letter to the editor by J. H[enry].

3

The Principal Transit Instruments

"A Transit telescope is a telescope moveable about an horizontal axis, and so adjusted as to make its line of collimation describe a great circle passing through the pole and zenith, or the meridian of the place. Its use is to take the right ascensions of the heavenly bodies, and to correct the going of the clock; for which purpose it has the system of wires described in art. 62. placed in the principal focus of the object glass. It has only one eye glass which is convex, and consequently the objects are inverted. Directing therefore the telescope towards the south, the bodies enter the field of view on the west side."[1]

So wrote the Rev. Samuel Vince, Professor of Astronomy at Cambridge, in 1790 some 100 years after the invention of what is today known as a transit instrument—an invention ascribed to Olaus Römer (1644–1710), the Danish astronomer who first used such an instrument near Copenhagen in 1689.

As we have seen in the last chapter, it was theoretically possible to obtain both RA and Dec from the same instrument, as Flamsteed had done with his mural arc. However, there were often good practical reasons for splitting the two functions. Halley used a transit instrument at Greenwich because he had no graduated mural instrument; Bradley and his successors used one because it became impossible to provide a single instrument which could be adjusted to provide both RA and Dec to the required precision, as was achieved later with the transit circle. If the instrument is perfect and set up accurately in the meridian, the star to be observed will be in the centre of the field of view at a sidereal time corresponding to the right ascension of the star. If the observed clock time of transit is

31

compared with the tabulated right ascension the difference gives the error of the clock at the time of observation. Similar considerations apply if the Sun or a planet is observed. The transit instrument, however, cannot be adjusted perfectly and three corrections must be applied.

If the pivots are not due east and west, but slightly north of east and south of west, stars north of the zenith will be seen after the time of transit, while those south will be observed to transit early. By observing stars both north and south of the zenith, and plotting the results on a graph, these errors, known as azimuth errors, can be detected and the appropriate corrections applied. If the east pivot is lower than the west pivot, the telescope is tilted towards the east and all stars, north or south, will be observed to transit early. This level error is measured by means of a spirit level which is placed astride the pivots or by nadir observations in a bowl of mercury.

The collimation error is caused when the optical axis of the telescope is not at right angles to the axis of the pivots; this is eliminated by reversing the telescope on its bearings.

3.1. *Halley's 5-foot Transit Instrument* (1721)
England's first transit instrument

Telescope said to have been made by Robert Hooke.
Mounted at Greenwich, 1721. Dismounted 1774.
Displayed in Old Royal Observatory since 1960.

Description (Fig. 31)

Telescope: brass tube 5 feet 6 inches long, aperture $1\frac{3}{4}$ inches, magnifying less than 40 times. Until 1742, only one vertical cross-wire. The object-glass survives, the eyepiece has disappeared. The object in Fig. 31 which looks like a handle (the wooden part is a recent addition) is probably a pointer for the setting semicircle fitted in 1746 which does not now survive.

Axis: 3 foot 6 inches long with telescope mounted 13 inches off-centre, supported by braces. This off-centre construction followed the practice of the instrument's inventor, Olaus Römer, but proved unsatisfactory in practice as the whole telescope and axis were liable to be affected when a hand was placed on them or when they were exposed to the Sun's rays.

Piers: It seems likely that the transit was supported by a free-standing pier on the east side and by the outer wall on the west side.

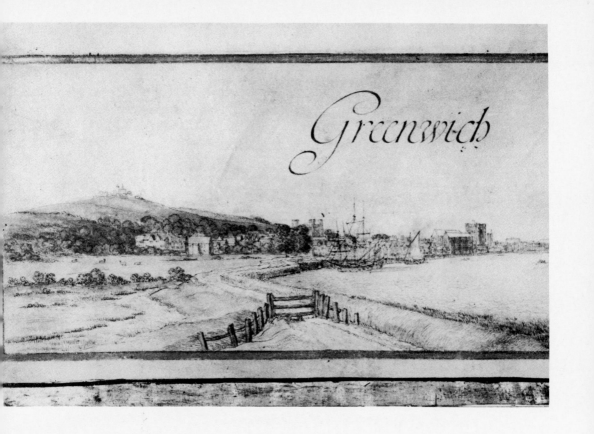

Fig. 1. GREENWICH CASTLE IN 1662
From a manuscript drawing attributed to Wenceslaus Hollar, on
Jonas Moore's plan of the River Thames. Reproduced by permission
of the London Museum.

PROSPECTUS INTRA CAMERAM STELLATAM.

Fig. 2. THE GREAT ROOM, ABOUT 1676

A and *B*—The Tompion year-clocks. *C*—The third clock. *D*—Flamsteed's 3-foot quadrant for checking the going of the clocks. *F*—Stand for telescope, adjustable in height. *SS*—Ladder for supporting telescope tubes, movable

portraits above the door were of Charles II and his brother James, Duke of York, painted by Lely or one of his school (present whereabouts unknown) From an etching by Francis Place after Robert Thacker, are

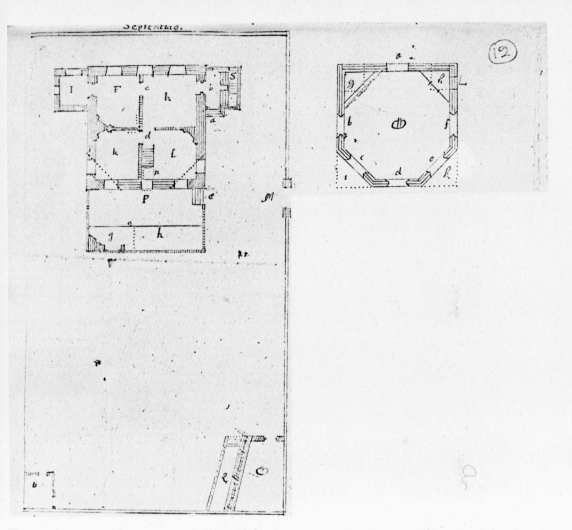

Fig. 3. THE ROYAL OBSERVATORY IN JANUARY 1676
Plans drawn by Flamsteed, before the building completion, described in his own words:

(a) This represents the figure of the middle story with the little yard & shop belonging to the neather *(sic)* & its position on the ground compassed with the outer wall represented by the outer lines. *M* the entrance at the great gates. *a* 3 steps rising into the entry or staircase. *b* the entry into the hall or room *h*. *c* the door of the room *F* designed for myself. *I* the study to it. *d* a door leading either into the rooms *k* or *I* or down the stairs into the lowest story. *e* 5 or 6 steps going into the paved yard *P* in which at *p* is a passage leading either on the right hand into the kitchen underneath *k* or a parlour without a chimney *l* or up the stairs into the middle story. *g* a small wash house with an oven on the corner. *h* a workshop. *S* the staircase leading up into the large observatory. *O* the watch-house or a room for the large sextans *(sic)* with a movable roof, but not yet fitted on the western wall cast as near as we could into the meridian. *Q* the side of it designed for a large semicircle or quad. *v* the house of ease.

(b) The room above the four chambers being 20 foot high and more than 30 foot wide is fashioned into an octagon *O* in which at *a b c d e f* are windows 4 foot wide but with mullions in the middle to sustain the glass so that only two foot is made to be opened on one side of each window for putting out the tubes. At *c* and *e* are doors in the windows to pass into the balconies *i* and *k* which are compassed with rails and ballisters and will be very convenient for hinging quadrants to take altitudes or a portable instrument for distances less than 10 degrees. Over the cant *h* is a lodging room to be for a servant and in the cant *g* are the winding stairs that lead to the roof, on which over *g* and *h* are two square turrets covered with lead and furnished with balls that add no little ornament to the building. The wings *I* and *S* in the ground plot reach no higher than the sole of the windows of this story so that they can hinder the prospect but little. The roof of this is almost flat leaded and compassed with rails and ballisters. I had sent you a draught of the front but I cannot find the design and therefore omit it and the prospect till a better opportunity.

Enclosed in a letter from Flamsteed to Richard Towneley 22 January 1676. (RSMS 243(re) Lr. 12) Reproduced by permission of the Royal Society.

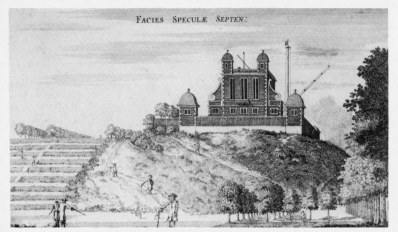

Fig. 4. THE NORTH FACE OF THE OBSERVATORY, ABOUT 1676
The Summer houses at the ends of the terrace, omitted from Fig. 3, may
have been architectural afterthoughts. Etching by Francis Place.

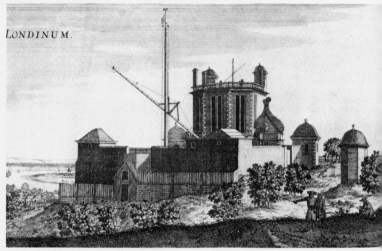

Fig. 5. VIEW TOWARDS LONDON, ABOUT 1676
The cupola on the right is the top of the well telescope, cupola to the
left of it the roof of the solar observatory (see Figs. 52 and 102). The
Sextant and Quadrant Houses can be seen here immediately under the
tall windows of the Great Room. Etching by Francis Place.

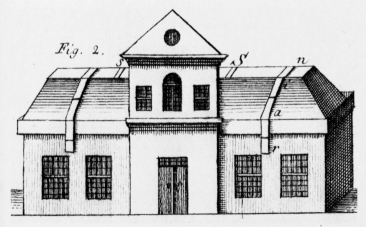

Fig. 6. BRADLEY'S NEW OBSERVATORY, 1769
Taken from a book by Frederick the Great'
royal astronomer describing his visit to Englan
in 1769, this is the only known picture of Bradley
'New Observatory', showing the hip-roo
which were removed in 1779 when gables wer
made on the east and west walls. This is a view
looking from the present courtyard. The Transi
Room is to the left, the Quadrant Room to th
right, the assistant's bedroom on the first floor an
the calculating room and library below. Fror
J. Bernoulli, *Lettres Astronomiques* (Berlin, 1771
Fig. 2.

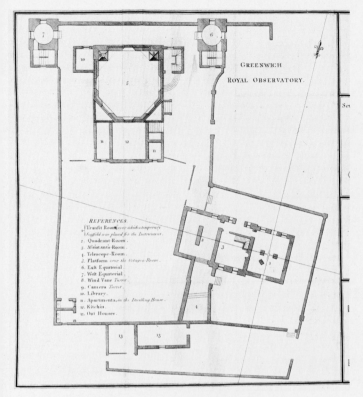

Fig. 7. PLAN OF 1788

Taken from General Roy's report of the Triangulation carried out
in 1787 to connect the observatories of Greenwich and Paris when
he erected his great theodolite precisely over the transit instrument.
This represents the state of the observatory generally from about
1750 until 1808 when the Circle Room was built onto the end
of the Transit Room (1). From *Phil. Trans.*, LXXX, plate XII.

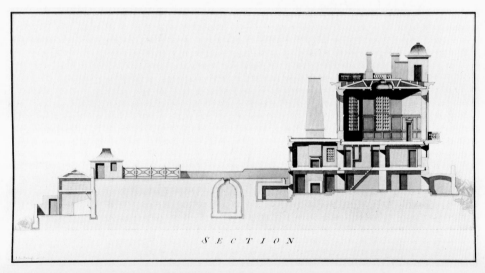

Fig. 8. FLAMSTEED HOUSE, ABOUT 1760

showing the tall chimney above the new living-rooms built, probably
about 1755, for Bradley. Coloured drawing by John Evelegh, with
two plans, presented to the Royal Observatory by Trinity College,
Cambridge, 1918.

Fig. 9. LOOKING WEST, 1794
A note on the back of this water-colour in the hand of Maskelyne's
daughter, Margaret, reads: "Royal Observatory Greenwich as it
was in 1794 drawn by Mr Sinley". The dome on the right contained
Sisson's equatorial sector, Wren's original summer house having been
converted in 1773. The present courtyard was enclosed not long
after this drawing was made. Water-colour in possession of N.
Arnold Forster, Esq., reproduced by permission.

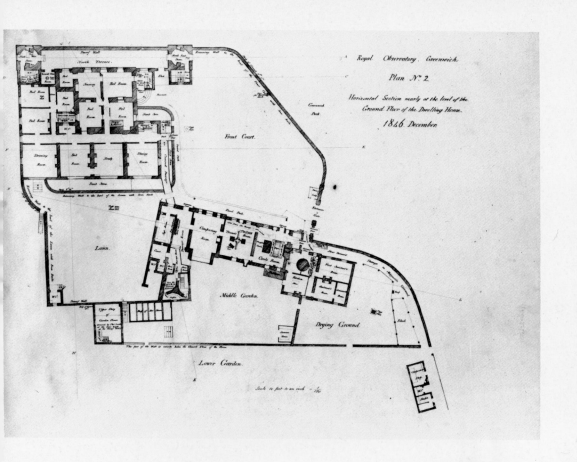

Fig. 10. PLAN, 1846
The three-rayed pier for Airy's new altazimuth can be seen at bottom
left. The Circle Room was to be converted into the Transit Circle
Room two years after this. Plans in possession of RGO.

Fig. 11. AIRY CROSSING THE FRONT COURT, 1839
Note the Gothic window of the Circle Room, removed when the
room was converted into the Transit Circle Room. The Sheepshanks
dome surmounts the East Building. The roof shutters from right
to left were for the 10-foot transit instrument (just to Airy's left),
the Troughton circle, and the Jones circle. Drawing in possession
of RGO.

Fig. 12. FROM THE ROOF OF FLAMSTEED HOUSE, LOOKING S.E.,
1869
Photo dated 3 July 1869, taken from the platform of the Robinson
anemometer *Left Centre*—drum dome for Great Equatorial with
Sheepshanks dome in front. *Right foreground*—drum dome for Airy's
altazimuth. *Centre background*—electrometer mast near the Magnet
House.

Fig. 13. LOOKING NORTH, ABOUT 1895
In the left foreground is the Lassell dome. In the centre is the Magnetic
Observatory with the 1881 Library to the right and the Onion Dome
of the 28-inch refractor (S.E. equatorial) behind. In front of Flamsteed
House is the drum-dome of Airy's altazimuth with the Astrographic
dome just to the left of the time-ball.

Fig. 14. THE SKYLINE ABOUT 1900
The domes from left to right are: Thompson, S.E. Equatorial, Sheepshanks, Astrographic, North-west (disused). The platform to the right
of the time-ball carried the Robinson anemometer.

Fig. 15. LOOKING WEST, ABOUT 1925
Wireless aerials can be seen leading into the Wireless Room (under the Sheepshanks dome) where foreign time-signals were received. The photoheliograph was installed in the old altazimuth drum-dome in 1911. The Astrographic dome can be seen at centre, with one of the transit circle shutters open underneath.

Fig. 16. H.M. QUEEN MARY IN FRONT OF THE NEW OBSERVATORY, ON THE 250TH ANNIVERSARY, 1925
The New Observatory was built by Christie to the south of the Old Magnet House in 1891–5. It was first called the New Physical Observatory, today it is the South Building. The dome housed the Thompson photographic equatorial, with a 26-inch refractor and a 30-inch reflector on the same mounting. Today it houses the Greenwich planetarium.

Fig. 17. WAR DAMAGE, 1940
The Altazimuth Pavilion in the foreground received considerable
bomb damage in October 1940. The coverings of other domes were
also affected.

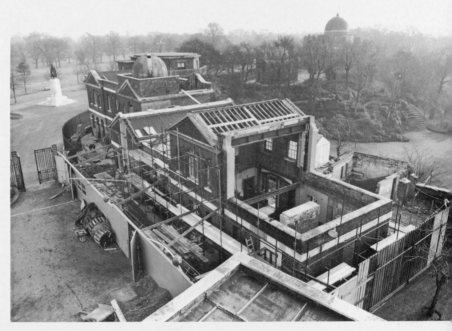

Fig. 18. RESTORATION, 1966
under the superintendence of the Ministry of Public Building and
Works, in preparation for the Meridian Building being opened to
the public as part of the National Maritime Museum. Taken from
much the same viewpoint as Fig. 14. Halley's quadrant wall can
be seen lower right. The South Building is at right background,
General Wolfe's statue is at left.

Quadrans Muralis Merid: 10 pedum Rad:.

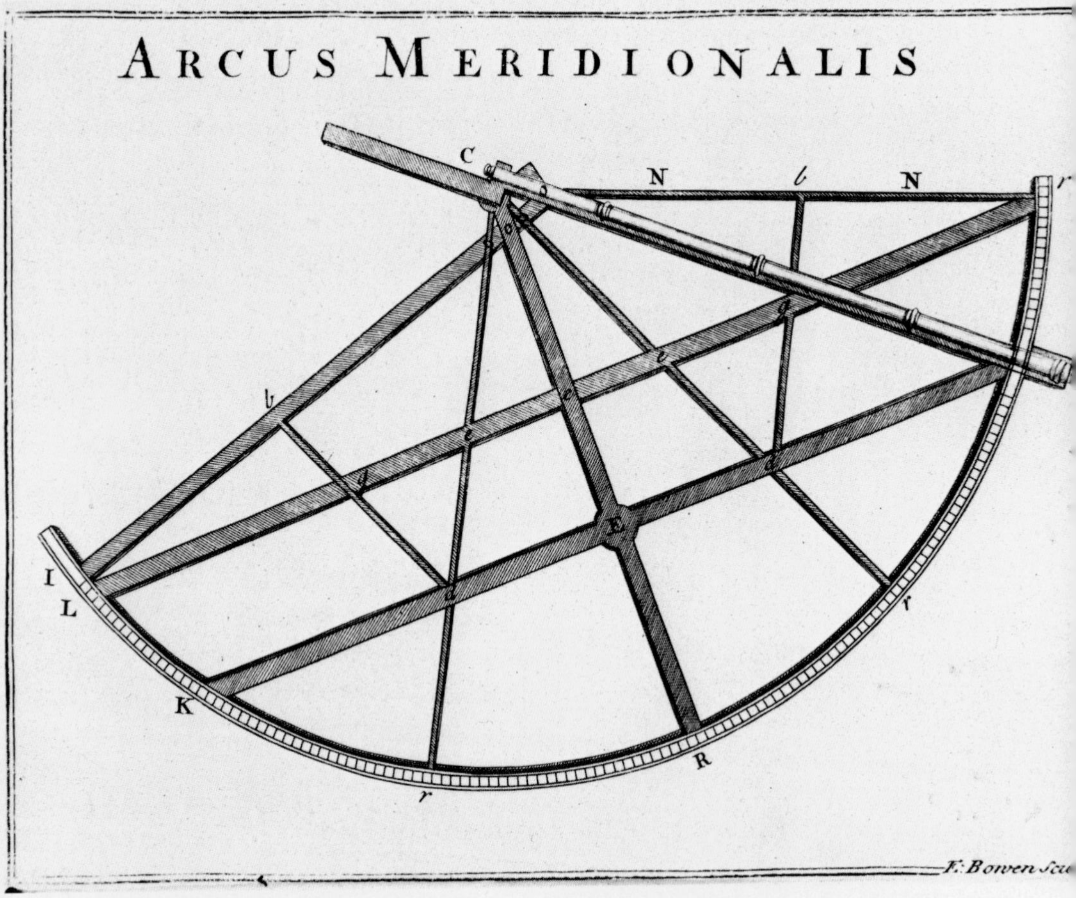

ARCUS MERIDIONALIS

F. Bowen Sc.

Fig. 20. FLAMSTEED'S 7-FOOT MURAL ARC, 1689
From Flamsteed's *Hist. Cel. Brit.* (1725).

ANN. CHR. 1698. Mense Die Styl. Vet.	Tempora per Horologium oscillatorium h ′ ″	Tempora vera apparentia. h ′ ″	ANNO MDCXCVIII.		Dist. a Vertice numeratæ. per Lineas Diagonales. o ′ ″	per Strias Cochleæ. Revol.Cent.	Dist. a Vertice correctæ. Er. 07 30 ′ ″
☉ Martii 13	12 15 10½		Virginis	γ	51 22 10	1164 36	51 14 40
	16 56½		quæ sequitur γ		51 29 15	1167 10	51 21 45
	21 32	12 18 20	Mars intrat		51 59 00	1178 30	51 51 30
	24 14½	21 06	transit		51 58 55	1178 28	51 51 25
	27 04	23 52	exit		51 58 45	1178 25	51 51 15
☿ 16	0 01 52		Solis limbus remotus		49 20 05	1118 23	49 12 35
	02 39		primus transit, centro		49 04 15	1112 17	48 56 45
	03 42		centrum transit, proximo		48 48 25	1106 26	48 40 55
			Nubes.				
	9 31 31½		Leonis	α	38 10 05	865 38	38 02 35
	10 40 24½			φ	53 35 20	1214 64	53 27 50
				d	52 55 20	1199 56	52 48 00

A   B   C   D   E   F

Fig. 21. PUBLISHED MURAL-ARC RESULTS FOR 1698
From Flamsteed's *Hist. Cel. Brit.*, II, 346.

Fig. 22. FLAMSTEED, WESTON AND HIS MURAL ARC ON THE CEILING OF THE PAINTED HALL

n the south-east corner of the ceiling of the Painted
Hall of the Royal Naval College, Greenwich, can be
een this portrait of the first Astronomer Royal with
is assistant and chief instrument, the mural arc,
ketched from life by Sir James Thornhill in about 1710.

The essayist, dramatist, and politician Richard Steele,
visited the Painted Hall on 12 May 1714, and described
t as follows:

> On the other end of the Gallery, to the South, is our
> learned Mr Flamstead, Reg. Astron. Profess. with his
> ingenious Disciple Mr. Tho. Weston. In Mr Flamstead's
> Hand is a large Scroll of Paper, on which is drawn the great
> Eclipse of the Sun that will happen on April [blank] 1715;
> near him is an old Man with a Pendulum counting the

> Seconds of Time, as Mr. Flamstead makes his Observations
> with his great Mural Arch and Tube. . . . .

The scroll on the right and the clock on the left show
the date and time of a total eclipse of the Sun, visible at
Greenwich, as predicted by Flamsteed in 1714. During
the restoration of the ceiling in 1960 it was found that
the time on the clock had at some time been changed
from 5.15 to 9.02 but there is no evidence when this was
done. The main painting had been completed by 1714
but the day of the month was left blank in Steel's account
of May 1714, quoted above. According to Flamsteed's
own account after the eclipse, it actually became total
at 9.09

Fig. 23. ONE OF BIRD'S 8-FOOT MURAL QUADRANTS, 1767

Graham's quadrant of 1725 served as the prototype for many quadrants made for observatories all over Europe by John Bird and Jeremiah Sisson between 1750 and 1790. This picture from a French encyclopaedia shows one of these British exports. Metal braces to prevent the telescope bending (seen in Fig. 25) have been omitted by the draughtsman. From the engraving *Quart de cercle Mural en Perspective et developement du contrepoid de la Lunette* by Benard, after Goussier, Plate X in Plates, Vol. V (1767) of the *Encyclopedie des Sciences, des Arts et Métiers*. The written description is contained in Volume XIII, pp. 667–671.

DAY of the MONTH.	Names or Characters of the Sun, Planets, and fixed Stars.	ZENITH DISTANCES								BARO-METER	THERMO-METER.	
		By the exterior Divisions.			By the exterior Divif. reduced.			By the interior Divifions.			With-in.	With-out.
		Div.	Part.	Vern.	Sec.	D.	M.	S.	D.	M.	S.	Inches.	Deg.	Deg.
☉ 16.	7 ⅞ { Leonis.	38	6	12	3	36	1	16,8						
		36	3	9	2	33	57	33,5						
	Regulus.	40	14	1	10	38	19	36,3						
	Jupiter's Center.	78	11	5	7	73	47	23,2	73	47	22	29, 70	38	31
☽ 17.	☉ Upper L.	43	5	12	0	40	38	58				29, 63	42	47
	☉ Lower L. *10	43	14	13	3½	41	10	53	41	11*	54½			
♂ 18.	Syrius.	72	5	14	5	67	50	44,3				29, 54	46	52
	Castor.	20	5	10	0—	19	4	46,5						

In MAY MDCCLVIII.

A   B   C

Fig. 24. PUBLISHED QUADRANT RESULTS FOR 1763
From Bradley, *Astron. Obs.*, II, 265.

 25. BIRD'S 8-FOOT BRASS MURAL QUADRANT, 1967
unted today in its 1750 position, this quadrant lacks the counter-
se on top of the pier and the associated wooden framework on
telescope which can be seen in Fig. 23.

Fig. 26. THE EYE-END
Bird's signature can be seen on the brass limb
above 45° on the inner scale, 48 units on the
outer. The eyepiece and attached double
vernier can be clamped to the limb wherever
needed. The micrometer can be seen under
the clamp to the left.

Fig. 27. JOHN BIRD (1709–76), INSTRUMENT MAKER

John Bird of London, who furnished the Chief Observatories of the World, with the most Capital Astronomical Instruments divided by him after an improved method of his own, in a manner superior to any executed before, for which, and many other Improvements in the Construction of Astronomical Instruments, he was honoured with a considerable Recompence from the Commissioners of Longitude.

(Title of the print of which this is a photograph)

The instrument shown on the engraving on the table is the Greenwich Brass Mural Quadrant, described in Bird's *Method of Constructing Mural Quadrants*, published in 1768, also seen on the table. As a result of this pamphlet and of the success of the Greenwich quadrant, similar instruments were made by Bird for the observatories at St Petersburg, Cadiz, Paris (two) and Oxford (two).

Smaller versions of the same design were scattered over Europe and surpassed in accuracy those of French and German manufacture. A Bird 6-foot quadrant was used by Tobias Mayer at Gottingen: Mayer's solar and lunar tables made possible the publication of the British *Nautical Almanac* in 1767, and were thus indirectly responsible for the choice of Greenwich as the Prime Meridian. He modified and repaired most of the instruments by Graham already at Greenwich, and, in addition, provided the following new ones: Bradley's 8-foot Brass Mural Quadrant, 1750. Bradley's 8-foot Transit Instrument, 1750. 40-inch Movable Quadrant, 1750. 20-foot Refracting Telescope, 1750. 2-foot Gregorian Reflecting Telescope. From a mezzotint by V. Green, 1776, after C. Lewis.

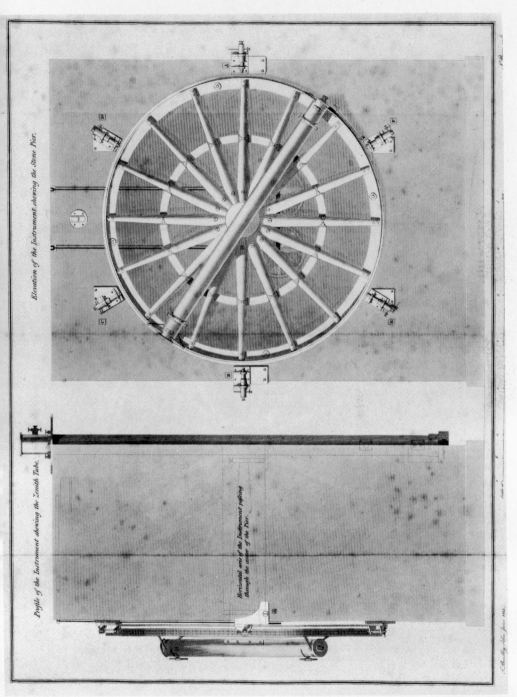

Profile of the Instrument shewing the Zenith Tube.

Horizontal axis of the Instrument resting through the center of the Pier.

Elevation of the Instrument shewing the Stone Pier.

Fig. 28. TROUGHTON'S 6-FOOT MURAL CIRCLE OF 1810
The telescope was firmly clamped to the "wheel", on the rim of which was the scale, the whole rotating on the axis. The "wheel" can in turn be clamped to the pier in whatever position is needed. The clamp can be seen, bottom right, near the eye-end of the telescope. The NPD is read through the inward-facing micrometer-microscopes 'A' to 'F' attached to the pier. From Pond, *Astron. Obs.*, I, 1811–13, frontispiece.

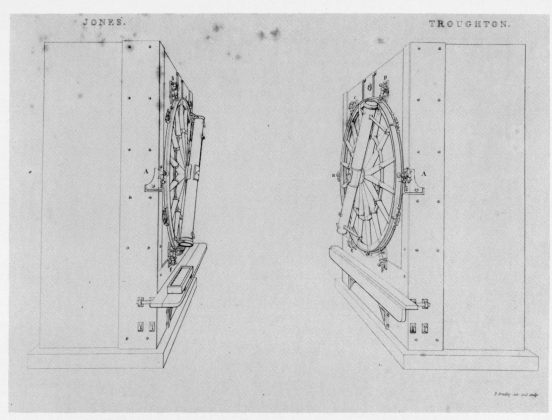

Fig. 29. THE TWO MURAL CIRCLES FACE EACH OTHER
A circle by Thomas Jones was erected opposite that of Troughton in 1825, permitting simultaneous observations—direct and by reflection from a bowl of mercury, as seen here for Troughton and Jones respectively. From Pond, *Astron. Obs.*, 1832, part V, Plate 1.

In the Year M.DCCC.XXXIII.

TROUGHTON.

	DAY of the MONTH.	Barom\(^r\)	Thermom\(^r\) In.	Out.	Names of STARS.	MICROSCOPES A	B	C	D	E	F	Mean of 6 Microscopes.	Mean of 2 Mic.	Observers.
1	♃ AUG. 15	29, 68	56	57	♀ - - - - - -	7 9, 6	9, 2	6, 6	5, 6	11, 5	15, 4	69 7 9, 7	9, 4	R.
2	♀ 16	29, 78	58	67	α Herculis - - - -	23 50, 2	47, 6	42, 3	44, 8	50, 1	53, 8	75 23 48, 1	48, 9	H.
3		..	..	...	α Ophiuchi - - R.	44 51, 8	47, 3	44, 2	50, 0	53, 4	56, 2	179 44 50, 5	49, 6	,,
4		....	57	54	γ Draconis - - -	28 54, 0	54, 7	51, 1	55, 6	58, 0	0, 2	38 28 55, 6	54, 4	,,
5		29, 80	..	53	*........ B. 170 -	31 56, 8	56, 8	54, 8	59, 3	0, 5	2, 4	38 31 58, 4	56, 8	,,
6		...	..	..	Next division.	57, 9	57, 0	55, 1	58, 2	0, 6	4, 3	58, 8	57, 4	,,
7		....	..	..	α Lyræ - - - -	21 28, 0	25, 7	24, 3	27, 7	28, 8	34, 3	51 21 28, 1	26, 8	,,

Fig. 30. PUBLISHED CIRCLE RESULTS FOR 1833
From Pond, *Astron. Obs.*,1833 (3), 16.

Fig. 31. HALLEY'S 5-FOOT TRANSIT INSTRUMENT OF 1721
The Ys on which it is mounted are not original. The 'handle' at
the right-hand end of the axis is in fact a pointer for a setting-
semi-circle once fitted.

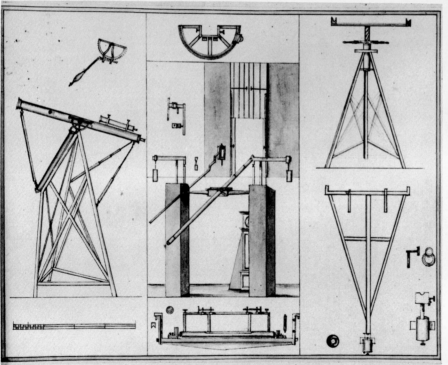

Fig. 32. DOLLOND 46-INCH ACHROMATIC TELESCOPE
(left) AND BRADLEY'S 8-FOOT TRANSIT INSTRUMENT
(right), SKETCHED ABOUT 1785
This is one of a series of drawings of the Greenwich instruments done from life about 1785 by John Charnock
(1756–1807) of Blackheath, known for his books
Biographia Navalis, History of Naval Architecture, and
Life of Lord Nelson. The drawings are preserved in the
NMM. Left—46-inch achromatic telescope by Dollond.

Centre—Transit room: the counterpoise apparatus and
the lamp arm have not survived. Centre top—setting
semi-circle (see Fig. 33). Centre bottom—Hanging spirit
level by Edward Nairne, 1774. Not survived. Top right—
Apparatus for reversing the axis—in effect, a jack. Not
survived. Bottom right—Plumb-line level by Jeremiah
Sisson, 1778. Not survived. From 'Charnock's Views',
Vol. IV, in NMM Print Room.

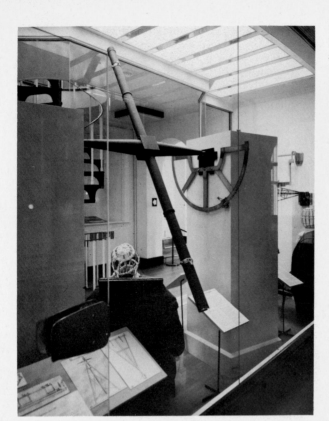

Fig. 33. BRADLEY'S 8-FOOT TRANSIT TODAY
Looking S.W. It lacks the counterpoises, the lamp
and the lamp arm, and the connection between
the axis and the setting semi-circle.

Obſerved TRANSITS of the FIXED STARS and PLANETS over the MERIDIAN.   5

In the YEAR MDCCLXXVI.

DAY of the MONTH.	1ſt Wire. M S	2d Wire. M S	3d or Meridian Wire. H M S	4th Wire. M S	5th Wire. M S	Names or Characters of the Stars and Planets.
☽ SEPT. 30	59,1	31,3	14 0 4,1	36,6	9,3	Arcturus.
	40,8	15,1	15 19 49,6	24	58,2	α Cor. borealis.
	51	21,5	15 27 52,1	23	53,7	α Serpentis.
	12,6	46,3	16 10 20,5	54,2	28,3	Antares.
	8,1	39,2	17 19 10,5	41,6	13	α Ophiuchi.
	13	44	19 30 15	46	16,9	γ ⎫
	28	58,8	19 34 29,5	0,4	31	α ⎬ Aquilæ.
	55,8	26,5	19 38 57	27,5	58,3	β ⎭
	49,1	20,5	19 59 51,8	23	54,3	1 α ⎫ Capricorni.
	13	44	20 6 15,5	46,7	18	2 α ⎬
	54,5	25	21 48 55,5	26	56,5	α Aquarii.
	13,4	44,9	22 48 16,1	47,5	19	α Pegaſi.
☿ OCT. 1	The Pendulum vibrates 1° 26' on each Side.					
		42,4	20 28 25,5	8,2	. . . .	α Cygni.
	20,1	54,5	23 51 29	3,5	37,8	α Andromedæ.
	18,7	50	23 56 21,4	52,8	24,2	γ Pegaſi.
	A	B	C	D	E	

Fig. 34. PUBLISHED TRANSIT RESULTS FOR 1776
From Maskelyne, *Astron. Obs.*, Transits, 55.

Fig. 35. TROUGHTON'S 10-FOOT TRANSIT TODAY, IN ITS 1816
POSITION

Fig. 36. INSCRIPTIONS ON TROUGHTON'S TRANSIT INSTRUMENT
OF 1816

Fig. 37. THE EYE-END,
with bubble setting circles.

Fig. 38. AIRY'S TRANSIT CIRCLE,
defining the World's Prime Meridian since 1884.

Fig. 39. THE EYE-END BEFORE THE IMPERSONAL
MICROMETER WAS FITTED IN 1915

Fig. 40. ZD MICROSCOPES,
on outside of west pier. The bottom microscope
marked 'P' reads degrees of south zenith distance;
'A' to 'E' read minutes and seconds only.

Pricker

Paper-covered drum

ZD micrometer

Contacts for electrical recording

RA micrometer

TROUGHTON & SIMMS. LONDON

Transit wire drive

Transit wire drive

41. THE TRAVELLING-WIRE (OR 'IMPERSONAL') MICROMETER,
TED 1916

Fig. 42. WILLIAM SIMMS (1793–1860), INSTRUMENT MAKER
William Simms, who went into partnership with Edward Troughton
in 1826, was a great friend of George Airy, After Troughton's death
in 1836, the firm of Troughton & Simms was directed by a member
of the Simms' family for almost 100 years. The firm became Cooke,
Troughton & Simms Ltd. in 1920 and are now part of the Vickers
Group. The more important Greenwich instruments connected with
the name of Simms are: Great zenith sector (1833), Altazimuth
(1847), Transit circle (1851), Reflex zenith tube (1851), Great equatorial (1858), Water telescope (1871), Reversible transit circle (1937).
From a photograph in possession of Vickers Instruments, York.

MAJOR PLANETS

Universal Time	Observer	Limb	Right Ascension Observed	O - C	Limb	Declination Observed	O´ - C	Diameter Horizontal Observed	O - C	Vertical Observed	O - C
						VENUS					
1954 d h			h m s	s		o ′ ″	″	s	s	″	″
Jan. 4 12	AM		18 32 40.67	+.12		−23 33 38.99	−2.29	0.47	−.25	10.52	+0.60
5 12	GS		18 38 10.37	+.06		−23 31 15.41	+3.29	1.17	+.45	10.45	+0.55
Mar. 12 13	PG		0 7 38.51	+.15		− 0 33 50.37	+1.83	0.70	+.04	9.65	−0.35
30 13	GS	2	1 29 34.99	+.07	S	+ 8 31 51.40	−0.70	..	..	..	..
						MARS					
Mar. 27 5	GS		17 30 33.45	−.02		−23 2 38.17	+2.43	0.85	+.17	13.86	+4.50
						JUPITER					
Jan. 4 22	AM		5 10 26.34	−.11		+22 29 57.35	−0.45	3.40	+.02	45.71	+1.93
5 22	GS		5 9 57.92	−.16		+22 29 34.61	−0.39	3.59	+.21	44.94	+1.24
15 21	PG		5 5 49.28	−.09		+22 26 15.80	−2.30	3.36	+.06	45.31	+2.49
16 21	AM		5 5 28.28	−.06		+22 26 0.20	−2.10	3.24	−.06	42.77	+0.05
17 21	GS		5 5 8.01	−.04		+22 25 45.97	−1.33	3.41	+.11	43.94	+1.32
30 20	LS		5 1 56.62	−.11		+22 24 0.49	−0.41	3.09	−.09	42.25	+1.11
Feb. 8 20	PG	..	5 1 (8)	..		+22 24 35.81	−1.99	..	..	41.43	+1.41
13 19	AM		5 1 11.68	−.04		+22 25 37.57	−2.13	3.04	.00	40.00	+0.62
Mar. 2 18	GS		5 4 1.78	−.10		+22 32 36.66	−1.14	2.80	−.08	38.04	+0.78
8 18	GS		5 5 57.27	−.05		+22 36 9.22	−1.68	2.88	+.06	37.67	+1.13
10 18	GS		5 6 41.80	−.03		+22 37 25.87	−1.93	2.82	+.02	37.89	+1.59
11 18	LS		5 7 5.10	−.07		+22 38 6.00	−1.30	2.66	−.14	36.49	+0.29
30 17	GS		5 16 35.75	+.05		+22 51 57.95	−1.45	2.68	+.04	34.64	+0.46
						SATURN					
Feb. 24 4	PG		14 30 59.89	−.20		−17 9 54.01	+0.39	1.20	−.02	18.96	+2.98

ig. 43. **PUBLISHED TRANSIT–CIRCLE RESULTS FOR 1954,**
ecording the last regular observations before the instrument was
anded over to the care of the NMM—an observation of Jupiter
y Mr Gilbert Satterthwaite on 30 March 1954 at 17^h 16^m 35^s 75.

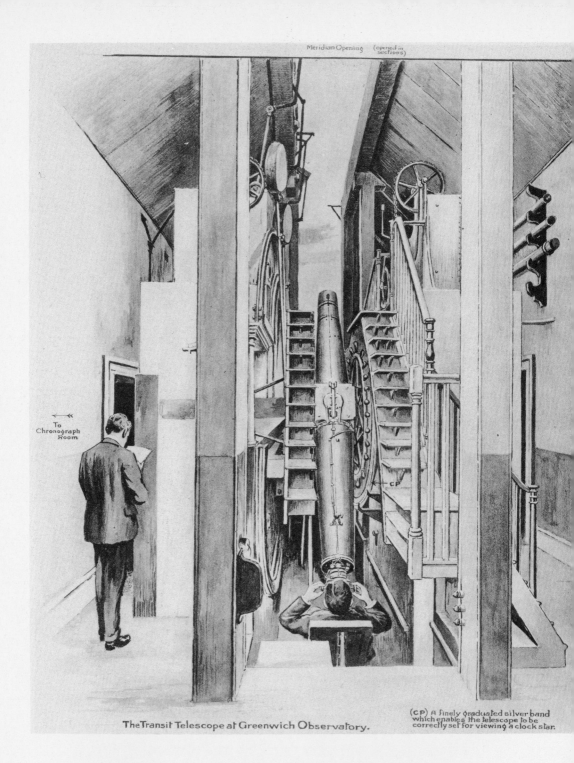

Meridian Opening (opened in sections)

To Chronograph Room

(CP) A finely graduated silver band which enables the telescope to be correctly set for viewing a clock star.

The Transit Telescope at Greenwich Observatory.

Fig. 44. AIRY'S TRANSIT CIRCLE IN 1923
From *The Illustrated London News*, 21 April 1923.

Fig. 45. THE COOKE REVERSIBLE TRANSIT CIRCLE AT GREENWICH
ABOUT 1940

Fig. 46. THE RTC AT HERSTMONCEUX, 1974

The 1726 Visitation said it was "substantially supported with free stones."[2]

Ys (the bearings upon which the ends of the axis rests): Nothing is known of the original design. The replicas now on display are based on Smith's *Opticks* (1738).

Counterpoises: These were fitted by Bradley.

History

We have seen in Chapter I how Halley, finding the Observatory devoid of instruments when he entered it in March 1720, obtained the promise of a grant of £500 for re-equipping. Though he was able to bridge the gap in the Great Room by using his own telescopes and a borrowed quadrant, he particularly lacked fundamental instruments and clocks. Furthermore, Flamsteed's meridian wall was believed to be subsiding through being too close to the brow of the hill.

We first learn of Halley's transit instrument in a letter dated 1 June, 1721 from Joseph Crosthwait, Flamsteed's last assistant, to Abraham Sharp, who did so much of the work on the mural arc:

> "He [Halley] has built a little boarded shed between the study and the summerhouse, and has fixed a stone in the ground, which stands about 4ft high: what he intends to fix upon it, I cannot yet learn: but as yet he has done nothing, neither has he anybody to assist him."[3]

What Halley had been doing was described in some detail after the Visitation of May 16 1726:

> "That he accordingly first provided a meridian instrument, being an axis fitted to an excellent 5ft telescope, made with very great care substantially supported with free stones, and placed in a little room built on purpose, adjoining to the west side of the Observatory and fitted up with proper openings both at the top and each end, for directing the telescope to all parts of the meridian.[4]

Later in the same report, the expenses are quoted.

The instrument itself and telescope	£30	0	0
The stone and mason's work	£1	10	0
A plain week clock to stand by it	£5	0	0
The building and fitting up the little room	£25	0	0
	£61	10	0[5]

The only near-contemporary account to mention a maker was Lalande, writing in 1764:

> "M. Halley se procura en 1721 une lunette de six pieds faite par M. Hook, mobile dans le méridien sur un axe, avec la quelle il commença à observer tous les jours la lune à son passage au méridien."[6]

Though Hooke himself had died in 1702, it is not impossible for the telescope at least to have been made by him. In 1720, Halley wanted instruments in a hurry. He could well have obtained a ready-made telescope, possibly getting Graham to fit the axis and braces—which look like afterthoughts.

The subsequent history may be summarized as follows:

1 October 1721: first recorded observation. 1725: after erection of 8-foot quadrant, regular observations with transit instrument ceased, and quadrant was used for both transits and zenith distances. 1730: last recorded observation with transit instrument by Halley (by then 74 years of age). 1742: death of Halley, James Bradley becomes Astronomer Royal. 1742: two extra cross-wires fitted by Sisson, counterpoises fitted, instrument balanced and adjusted by Bradley. January 1743: start of regular transit observations with this instrument, shorter focus eyepiece fitted. 1746–47: new setting-semicircle and level by John Bird. 31 August 1750: last observation with Halley's transit. 2 September 1750: first observation with Bradley's transit. 1774–79: Halley's Transit Room demolished; transit and level put into store.

1818: in Transit Room as a relic. 1850: to Transit Circle Room, west wall.

1953: on display in Octagon Room. 1960: in showcase in Halley Gallery, Flamsteed House, 1967: mounted in present position on replica piers.

Contemporary Accounts

1738: R. Smith, *A Compleat System of Opticks*, 1738, para, 841. 1771: Jerome Français de la Lande, *Astronomie*, 2nd Edition 1771, paras. 588 and 2388.

1836: S. P. Rigaud, "Remarks on Dr Halley's Instruments at Greenwich", *Mem. R. Astron. Soc.*, 9 (1836), pp. 205–214.

3.2. *Bradley's 8-foot Transit Instrument* (1750)

By John Bird of London.

Mounted in Transit Room, defining Greenwich meridian from 1750 to 1816.

Displayed on replica piers in Middle Room, Old Royal Observatory, from 1967.

Description (Figs. 32 and 33)

Telescope: originally, common OG, 8-foot focal length, aperture 2.7 inches stopped down to 1.5 inches by elliptical reflector. Fixed eyepiece. Magnitude 50. 1765: eyepiece made capable of focusing. 1772: achromatic OG by Peter Dollond, 8-foot 1-inch focal length, aperture 2.7 inches. Single positive-focus eyepiece capable of being moved laterally to view each of the five wires separately. Over the years there were various other eyepiece changes of less importance.

Cross-wires and Illumination

Five vertical and one horizontal wire until 1784 when two horizontal wires were fitted. Silver wire 1/750 inch thick, later 1/1000 inch. Illumination originally by lamp slung on pole attached to the East Pier which could be raised or lowered from the eye-end of the telescope: light reflected down tube by annular reflector in front of OG. Maskelyne used "solid elliptical illuminators, placed in the axis of the telescope."[7] (This is not understood as there seem to be no apertures in the axis through which light could be thrown.)

Axis

Exactly 4 feet long.

Counterpoises

The apparatus seen in Fig. 32 on top of the piers lifted the ends of the axes almost off their bearings. The total weight was 55 lb. Bradley set the counterpoises so that only 3 lb were bearing.[8] Bernoulli stated in 1769 they were set to give a weight of less than 1 lb.

Ys

Adjustment for cross-level on east pier, for azimuth on west pier; latter operated when needed from eye-end while viewing meridian marks, using a long pole with a Hooke's joint. Block for eastern Y, 8 × 8 inches; for western Y, 9 × 8 inches.

Some years after the instrument was dismounted in 1817, its Ys were mounted (in 1831) in the northern aperture of the Transit Room to support the 5-foot transit instrument used as collimator for the 10-foot instrument; they have not survived. Those supporting the instrument since 1967 were originally part of the Bird transit from Oxford, the locating pins and bolt holes of which exactly fitted the Greenwich setting-semicircle. Thus, the Greenwich *Ys* must have been almost identical to the Oxford ones.

Setting-semicircle, 20-inch radius, on west pier, graduated in declination to 10 minutes of arc, vernier reading to 2 minutes of arc. This has survived, though the connection between the vernier index and the telescope axis is missing.

Apparatus for Reversing the Axis

"An engine to be provided, consisting of an upright pillar and a strong upright screw for lifting the axis of the transit instrument off the notches in which it rests, and turning it half-round for the easier adjusting of the line of collimation . . .5 guineas"[9] (see Fig. 31).

Levels

1750: Hanging spirit level by Bird. 1772: Hanging spirit level by Edward Nairne (see Fig. 32). 1777: Hanging plumb-line level by Jeremiah Sisson (see Fig. 32).

1805: Striding level by Troughton, converted 1816 to work with 10-foot transit. Survives.

Piers

2 feet square, 6 feet 2 inches high. After being used also for the 10-foot transit, the piers were removed in 1851 and have not survived. The stone slab on which they were cemented is still in place under the floor and replica piers are now mounted on top.

Meridian Marks

South: iron plate with hole through which sky is seen on the house of Lord Chesterfield (later Mr Hulse's, now the Ranger's House).

North: on a house in Greenwich with a hole through which river is seen. Initially, these were brought to focus by placing a stop in front of OG. In 1778 and 1780, special subsidiary OGs of a quarter mile and three-quarters of mile focal length were obtained. In 1798 a new north mark was placed in the courtyard, 78 feet from the instrument and a special OG was provided for viewing it; the stone pier still survives and was used from 1920 to 1950 as the support for the night-sky camera.

Method of use (Fig. 34)

1. Set the telescope to approximate NPD of star to be observed, using index on setting-semicircle.

2. Set eyepiece to right-hand side of field of view, centred on wire 1, and wait for the star to enter field.
3. When star appears, adjust setting of telescope to bring image of star onto horizontal wire.
4. Note clock time of transit of star across wire 1 by listening to the beats of the clock, interpolating to a tenth of a second.
5. Record clock time of transit (column A, Fig. 34).
6. Move eyepiece until field centred on wire 2.
7. Record time of transit of star across wire 2 (column B).
8. Repeat procedure for wires 3, 4 and 5 (columns C, D and E).

Historical Summary

1749: made by John Bird for £73 16s. 6d.; erected in the Transit Room of Bradley's 'New Observatory'; transit clock 'Graham 3' (by Shelton under Graham's supervision), on south wall, east of shutters. 2 September 1750: first regular observation; 'Bradley Meridian' established through centre of instrument. 1765: Maskelyne became Astronomer Royal. 1767: *Nautical Almanac* first published, based on 'Bradley Meridian'. 1772: Dollond fitted achromatic object-glass and new sliding eyepiece; new hanging level by Edward Nairne. 1777: plumb-line level brought into use; Bugge reports white paper pasted directly on tube to reflect Sun's heat. 1779: to keep room cool and improve seeing conditions, roof openings widened from 6 inches to 3 feet; low ceilings removed; transit clock moved from south wall to new clock pier close to transit. 1784: mahogany covers made for top of piers to keep Sun off pivots. 1787: first London–Paris triangulation; General Roy's theodolite erected on roof of Transit Room (Fig. 7). 1798: new north meridian mark in courtyard.

1800: "Just before the Sun's transit found a little spider in the tube, beyond the wires, which had made itself a web there" (Maskelyne, *Astron. Obs.*, IV, 42); the spider was allowed to remain there from 24 June until 1 July. 1803: axis-pivots and Y's re-ground after 53 years of use, by Edward Troughton. 1807: 6-foot circle proposed, to take the place of both the transit instrument and the two quadrants. 1811: death of Maskelyne; John Pond became Astronomer Royal. 1812: mural circle by Troughton brought into use; it was hoped that this would serve to observe transits as well as NPDs but it did not prove sufficiently stable, so a new transit instrument was ordered from Edward Troughton in 1813. 5 July 1816: last observation with 8-foot transit, 10-foot transit erected on same piers; first observation, 21 July; old transit hung on wall of Transit Room. 1820: object-glass from old transit fitted into a new tube and used as a zenith telescope on the back of the Circle pier. 1850: tube, axis and setting-semicircle to Transit Circle Room, west wall.

1960: in showcase in Halley Gallery, Flamsteed House. 1967: mounted on replica piers in showcase in Bradley's Middle Room; the OG is now in 8-foot Zenith telescope in Bradley's Transit Room.

Contemporary Accounts
c. 1760: Bradley, *Astron. Obs.*, Introduction. 1776: Maskelyne, *Astron. Obs.*, I (1776), i–vi. 1777: Bugge diary (Copenhagen), 89–92 for detailed description and drawings of plumb-line level.

3.3. *Troughton 10-foot Transit Instrument* (1816)
By Edward Troughton of London. OG by Peter Dollond.
Mounted in Transit Room, defining Greenwich Meridian from 1816 to 1850.
Displayed on replica piers in original position in ORO since 1967.
Inscribed: [On one side of centre of axis] *Designed and Executed for the Royal Observatory by Edward Troughton, London, 1816. The optical part by Dollond.*

[On the other side] *To the President of the Council of the Royal Society this and the Mural Circle, being his greatest and best works, are dedicated by the maker.*

Description (Figs. 35–37)
Object-glass: Achromatic, 5-inch aperture, 9 feet 8 inches focal length. Made in 1793 by Peter Dollond (the largest glass he ever made) for an ordinary 'gazing' telescope for the Observatory. Transferred to Reflex Zenith Tube 1851 (now in Pond Gallery, ORO).
Eyepieces and cross-wires: One frame with five fixed vertical wires and two fixed horizontal wires, one with seven fixed vertical wires and two vertical wires movable by a micrometer. Magnifications between × 125 and × 500. Illuminated through axis by lamp placed on appropriate pier.

Axis
3 feet 6 inches long. Pivots of bell-metal until 1825, then steel.

Ys

Both with adjustment for level and azimuth. Those in place today are replicas.

Setting-circles: Originally 12-inch radius attached to axis, read by microscope on pier (not survived). Later, circles each side of eye-end of telescope with movable indices bearing spirit level.

Counterpoises: Removed 1816.

Piers: Lower part 2 feet square by 6 feet 2 inches high used originally for Bradley's transit; to accommodate longer telescope and shorter axis of 10-foot transit, semicircular caps, 2 feet diameter, 2 feet 3 inches long, were placed on top.

Levels: three striding levels, one of which was made in 1805 by Troughton for 8-foot transit and adapted for 10-foot transit in 1816.[2] All survive.

Reversing the axis: From 1840 Inventory:

"16. An iron crane attached to the tie-beam on the western side of the opening of the Transit-Room roof, turned upon its axis by a rope; for sustaining numbers 13 [the level] and 17 [see below]

17. A set of pulleys, leathern saddle, straps and buckles, for raising the transit instrument in reversion".

Mirror

"21. A looking-glass carried by hinges on the western pier, for shewing the clock face in observations of northern stars".

Meridian marks: 1816 (*a*) in courtyard 78 feet distant, (*b*) on wall of engine house in Blackwall. 1824: granite obelisk erected at Chingford, Essex, 11 miles away. 1834: Blackwall engine house demolished. Mark placed on new tavern 10,706 feet distant. No mention of south meridian mark.

Collimator: From 1831, 5-foot transit instrument mounted on *Ys* (once belonging to 8-foot transit) in northern aperture of transit room.

Historical Summary

1813: ordered from Troughton (cost £315) after it had been found that the new mural circle was not sufficiently stable to take accurate transits; instrument designed around 5-inch object-glass of 10-foot focus made for an ordinary telescope in 1793 by Peter Dollond. 1816: Bradley's old 8-foot transit instrument dismounted and new instrument mounted on same piers. Stone capping pieces were added to piers to raise height by 1 foot and reduce distance apart by 6 inches; first regular observation, 21 July; transit clock 'Graham 3'. 1825: transit clock 'Hardy'. 1825: new steel pivots by Troughton. 29 December 1850: last observation; superseded by Airy's Transit Circle. 1851: telescope and axis hung on west wall of Transit Circle Room; object-glass transferred to Airy's Reflex Zenith Tube.

1960: in Halley Gallery, Flamsteed House. 1967: re-erected in its 1816 position on replica piers and *Ys*; OG and RZT in John Pond Galley, ORO.

Contemporary Account

1829: W. Pearson, *An Introduction to Practical Astronomy*, II, (1829) 366–71.

3.4. *Five Small Reversible Transits* (1870)

Signed: *Troughton & Simms London 1870*

Five ('A' to 'E') purchased 1870 for Transit of Venus (see p. 119).

'A' remained at Cape Observatory.

'B' to 'E' used intermittently for longitude determination, 1882–1926.

'E' sold to Egypt, 1929.

'B' to 'D' used for time-determination at Greenwich and Abinger, 1927–57.

'C' to NMM, 1967.

Description (Fig. 111)

Telescope: 3-inch aperture. 36½-inch focal length. Impersonal micrometers fitted from 1921.

Axis: 18 inches long, exclusive of gunmetal pivots which are 1¼ inches long and 1½ inches diameter.

Mounting: At T of V, stone piers, 4 feet 11 inches high, weighing 1400 lb each, sent out from England. From 1892, cast-iron stands were provided, fitted with apparatus for rapidly reversing the telescope in its *Ys*.

Method of use

In the eyepiece of the instrument there is a vertical spider's thread which is moved across the field of view by means of a micrometer screw driven by hand. The observer moves the thread towards the east of the field and when the star image reaches the thread he bisects the image and turns the micrometer to continue to bisect the star image as it traverses the field. As the micrometer is turned, electrical contacts are made and these are automatically recorded on a chronograph tape; a second pen records pulses from a standard clock. The tape is measured and the clock time of each signal from the transit instrument is determined. A mean of ten signals before and ten signals after transit is used to determine the time of transit of each star.

In practice, an astronomical observation for the determination of the error of the clock consists of the records of clock times of transit of about 12 stars, preferably 6 north and 6 south of the zenith. The telescope is set to the correct declination for the required star and observing commences a minute or more before the tabulated right ascension (longer for a slow polar star); ten contacts are recorded before meridian transit, the instrument is reversed and ten contacts are recorded after meridian transit. A correction is made for the length of the contact. This is repeated for 12 stars.

Before observing commences, in the middle of the observation and after observing ceases, measurements of the level error are made using the spirit-level. If the spirit-level is placed astride the pivots, the telescope is set to a fixed altitude, about 70°, so that any error due to the shape of the pivots on top is systematic rather than random. Each observation of the level error consists of measurements of the bubble taken in four positions, with the telescope pointing north and the striding level in one east–west direction and then reversed, and with the telescope pointing south and the striding level in the two positions.

If the bubble is fixed on the cube of the instrument, the telescope is set pointing to the zenith and readings of the bubble are taken in pairs with the telescope reversed in between. In this case a minimum of six pairs of readings is taken.

Historical Summary

T of V: 1874 and 1882, 'A' to 'E'.

Longitude-determination, 'B' to 'E',—Greenwich to Paris, 1888, 1892 and 1902; Montreal, 1892; Killorglin (Ireland), 1898; Malta, 1909; various 1926.

Time-determination at Greenwich (in Transit Pavilion on Bradley's meridian in courtyard): 1927, 'B'; 1933, 'D'; 1935, 'B'; 1937, 'C'; 1937, 'B'.

Time-determination at Abinger: 1940, 'D'; 1940, 'B'; 1943, Bamberg (p. 120); 1946, 'D'; 1947, Bamberg.

Time-determination at Greenwich (in Transit Pavilion and Altazimuth Pavilion): 1946, 'B'; 1953, 'C'.

Time-determination at Herstmonceux: 1957, 'C'; superseded by PZT October 1957.

Contemporary Accounts

1874: G. B. Airy (Ed.), *Account of the Observations of the Transit of Venus, 1874—the British Observations,* (1881), 9–10. 1888–1902: W. H. M. Christie (Ed.), *Royal Observatory, Greenwich—Determinations of Longitude, 1888–1902* (Edinburgh, 1906), 6–8, 116–7, *5, 127.*

NOTES

1. Vince, p. 75.
2. ROV Vol. I, 21.
3. Baily, p. 344.
4. ROV Vol. I, 21.
5. ROV Vol. I, 23.
6. Lalande, paras. 588 and 2388.
7. Maskelyne, *Astron. Obs.,* I (1776), iii.
8. Bradley, *Astron. Obs.* I (1798), iv.
9. ROV Vol. I, 156.

4

The Transit Circles

THE transit circle is designed to yield both RA and ZD (or Dec) on the same meridian transit. The first instrument of this kind was probably the meridian circle which Horrebow erected near Copenhagen before 1735. Cary made one for the Rev. Francis Wallaston in 1793 but the axis proved too slender and the circle too small. The first really successful transit circle was made by Troughton in 1806—for Groombridge of Blackheath, later passing to Sir James South.

At Greenwich, Airy's transit circle of 1850 superseded both Troughton's transit instrument and his mural circle.

4.1. *Airy's Transit Circle* (1850)
Defining the World's Prime Meridian since 1884

By Ransomes & May of Ipswich (engineering) and Troughton & Simms of London (optical and instrumental) to the design of G. B. Airy.

Mounted in Transit Circle Room, defining Greenwich Meridian since 1851. Still in working order, 1975.

Signed twice on telescope tube: *RANSOMES & MAY ENGINEERS IPSWICH 1850.*

Impersonal micrometer (1915) signed: *TROUGHTON & SIMMS LONDON.*

Description (Figs. 37–44, 126)

Transit telescope: OG 8·1 inches aperture, 11 feet 7 inches focal length. Magnification 195 (180 for Sun).

Eye-end: originally seven vertical fixed wires and one horizontal wire movable by micrometer; six additional vertical wires added 1854 with introduction of chronograph to give nine wires for galvanic registration and seven for 'eye-and-ear': in 1891 a ten-wire galvanic system was introduced to simplify arithmetic: travelling wire micrometer described in Fig. 41 fitted 1915. Magnification normally 195 but 180 for Sun: Bohenberger nadir eyepiece for ascertaining level errors.

Axis of cast iron, 6 feet overall. The pivots are of 'chilled' iron 6 inches in diameter. Telescope and axis approximately 1890 lb, reduced by counterpoises to about 150 lb on each Y.

Vertical circle: 6 feet diameter cast-iron, fixed to axis on west side: on outer surface of circle is a silver band with main graduations, very accurately divided from 0° to 360° with 5' spaces read by ZD microscopes pointing through west pier. On inner side of main circle is a setting circle, roughly divided to 5' with two pointers, one for ZD another for NPD.

ZD microscopes (see Fig. 40): pointing at inner circle through individual perforations in west pier, circle being lit by lamp through additional perforations in pier. Brilliant illumination of the circle by specular reflection when viewed through microscope was achieved by setting the silver band at a precise angle to the cast-iron circle. Pointer-microscope P reads a few minutes less than south zenith distance (0° = zenith, 90° = south horizon, 180° = nadir, 270° = north horizon). Micrometer-microscopes A, B, C, D, E, F at 60° intervals give minutes and seconds reading to 1/1000 of a minute; a, b, alpha and beta are for test purposes.

Piers: the east pier, of granite, had supported Troughton's mural circle from 1812 to 1848. The west pier, of Portland stone, now pierced for ZD micrometers, had supported Jones's mural circle from 1824 to 1848. Using different materials for the one instrument was to cause much trouble in the future, due to seasonal variation of level and azimuth errors.

Mercury trough: for observations by reflection and for nadir observations for level error. Can be moved as convenient or swung out of the way when not in use.

Sun shield: light alloy and cardboard screen about 3 feet long and 2 feet wide replaces flat brass ring on dew cap during Sun observations. The sunlight passes through the central hole 8 inches in diameter, the tube and axis being shielded.

Steps: primarily for the observer making mercury-trough observations with the telescope pointing below the horizontal: they
also screen the vertical circle from the Sun's rays around noon.

Collimating telescopes (two): originally 4 inches aperture, 5-foot
focal length. Present telescopes (1865–66) 7-inch aperture, 6 feet
10 inches focal length.

Roof shutters: 3 feet width opening. An example of the care Airy
lavished upon the smallest detail of the whole design occurs in the
roof shutter winches. Just before the shutters reach the fully closed
or fully opened position, a hammer strikes a bell "as a warning to the
person who turns the winch that he must move it gently".*

Method of use (Fig. 43)

Observing a star more than 5° from Pole, assuming Airy barrel
chronograph (which ran continuously) was in use. There were generally two observers, one on the telescope, the other on the circle
microscopes.

PREPARATION 1. An equatorial star moves over one wire interval
in about 15 seconds. At a time $15(1+\sec \delta)$
seconds before transit, set telescope at approximate
North Polar Distance of the star, using setting
circle and pointer.

2. Look into eyepiece and set movable transit wire
slightly to right of fixed wire 3½. Star should be
seen approaching wire 3.

3. Adjust setting of telescope in NPD until star is
approximately bisected by horizontal wire.
Clamp telescope.

FIRST 4. Bisect star with movable horizontal wire, using
ZD ZD micrometer-head to move wire and its frame.
OBSERVATION Make a 'prick' on the paper strip covering
micrometer drum at some point in the field
between wires 3 and 3½. Move wire away from
star and step pricker up by one notch.

RA 5. When star reaches movable RA wire, start turning
OBSERVATION RA screw by hand-wheels and maintain continuous bisection of star image from wire 3½ to
wire 4½. 18 micrometer taps will automatically
be recorded on chronograph, 3 per revolution of
micrometer, and of these 10 will be read-in later
against time scale.

*Airy, *Astron. Obs.* 1852, Appendix 4.

SECOND 6. Bisect star with movable horizontal wire between
ZD wires 4½ and 5 so that the mean of the 'pricks'
OBSERVATION corresponds with the meridian, wire 4. Step up
one more notch. If top of the paper is reached, turn
down 10 notches and advance drum by 1-tenth
of a revolution. Otherwise leave drum at position
of the last 'prick'.

7. Read circle microscopes and pointer and enter
into observing book: (a) RA and name of star,
(b) position of ZD observations, and (c) any other
notes.

8. Then or later, enter in note book (d) ZD drum
reading.

Notes

(1) *Close polar observations* were made differently owing to their
very slow speed. The star was not followed continuously. Ten
individual bisections were made, each time releasing the hand drive
and depressing a recording key simultaneously.

The ten micrometer readings were entered into the note-book
and were later related to the times of contact registered on the
chronograph.

The ZD observations consisted of two 'pricks' made at some
convenient point in the field just before or after the RA observation.
The accurate position in revolutions or fractions of a wire interval
was recorded.

(2) *For planets and full Moon:* at step 4, separate 'pricks' are made
for both north and south limbs; at step 5, transit wire is kept
tangential to preceding limb before transit, to following limb after
transit; at step 6, pricks are once more made for both limbs.

(3) *If tape chronograph was in use,* switch on after step 3, switch off
after step 6.

Historical Summary

In his report to the Board of Visitors in 1847, Airy said:

> "I think it worthy [of] the careful consideration of the Visitors,
> whether meridional instruments carrying larger telescopes should
> not be substituted for those which we possess. Whatever we do,
> we ought to do well. Our present instruments were, at the time of
> their erection, the best in the world; but they are not so now: and
> we actually feel this in our observations."*

*Airy, *Report*, 1847, 11.

In 1848 he persuaded the Admiralty to buy from Simms for £275 an object-glass of 8-inch aperture and 11 feet 6 inches focal length. The transit circle was designed around this glass, largely by Airy personally. The subsequent history may be summarized as follows: 1848–9: Circle Room converted into Transit Circle Room. 1850: transit circle erected 5½ feet south and 19 feet east of old transit instrument; transit clock 'Hardy' in niche at south end of pit. 4 January 1851: first observation; Airy had intended that this should be on the first day of the new half century but he was frustrated by the English weather. 27 March 1854: chronograph brought into use for recording observations; clock impulses from 'Hardy'. 1865–66: present collimator telescopes fitted; central cube of transit circle telescope pierced to permit collimating observations without raising telescope. 1870: sheet of very thin paper placed under eastern Y to compensate for sinking of pier. 1873: new apparatus fitted to ZD micrometer at eye-end to allow observations to be recorded by punching holes on a strip of paper, enabling observer to make several independent bisections of a star image without moving eye from the telescope. 1880: Greenwich Mean Time, based on meridian of this instrument, made legal time in Great Britain. 1882: collimating telescopes made to swing clear of meridian to allow reflection observations of stars to be taken as far as 71° on each side of the zenith. 1884: International Meridian Conference in Washington recommended the World's Prime Meridian should be the meridian passing through the centre of this instrument. 1891 and 1906: OG repolished by Troughton & Simms. 1908: illumination of circle and of field changes from gas to electricity. 1915: travelling-wire micrometer (impersonal micrometer) fitted. 1922: some divisions on circle having been obliterated by cleaning over the years, certain graduations were re-engraved *in situ*. 1927: superseded for time-determination observations by small reversible transit telescope set up in Transit Pavilion in courtyard. 1930: decision taken that new transit circle should be purchased. 1938: last major programme completed; new reversible transit circle now in operation. 1940: observations suspended due to wartime conditions. 1942: limited programme of observations restarted because of destruction of Pulkovo Observatory during the bombardment of Leningrad. 26 April 1944: artificial illumination of collimator wires installed, allowing collimation to be measured at any time of day. 23 March 1949: paper type chronograph installed, allowing measurement of transit registrations to $0^s \cdot 01$ and reduction of observations to $0^s \cdot 001$ for first time. 1951: 10-inch circular mercury bath replaced old trough; new bath made to sit on removable platform which rests on lugs driven into the base of the piers just below the dewcap when taking nadir observations. 1954: last regular observation after some 600,000 observations over 103 years. 1967: placed on view to the public; the instrument remains in working order and observations are taken periodically.

Contemporary Accounts

1852: G. B. Airy, 'Description of the Transit Circle ...', *Astron. Obs.*, Appendix, 1852 (reprinted 1867).

1952: W. M. Witchell "The Story of the Greenwich Transit Circle", *Occ. Notes R. Astron. Soc.*, 1952 December, 147–149.

4.2. *Cooke Reversible Transit Circle* (1933)

Signed: *Cooke Troughton & Simms Ltd, London & York, 1933.*
New objective by Cox, Hargreaves & Thompson, 1952.
Mounted at Greenwich, 1936–53.
Mounted at Herstmonceux, 1953.
Still in use, 1975.

Description (Figs. 45 and 46)

Unlike the Airy transit circle, it has a motor-driven impersonal micrometer and is reversible.

Transit telescope: Original objective, doublet 7-inch aperture, 8 feet 6 inches focal length. New coma-free objective (1952). 8 feet 6·5 inches focal length.

Eye-end: Motor-driven travelling vertical wire in impersonal micrometer. Speed of travel set by observer according to NPD: observer can adjust position of moving wire to keep wire on body.

Axis: 4 feet $3\frac{1}{2}$ inches overall. Two chains, attached to counter-balance arms supporting massive lead weights, reduce weight on each pivot bearing to less than 50 lb.

Vertical circles: Two, of glass, 28 inches diameter, with divisions etched on 24-inch annulus at 5' intervals, reading to $0' \cdot 001$ by microscope. In 1952 the eye-ends of the micrometers were replaced by cameras, seven for each circle (six circle-readers and one pointer).

Piers: Hollow, of cast iron, filled with anti-freeze mixture to counteract thermal expansion or contraction.

Collimating telescopes: 7-inch objectives identical with original objective of main telescope. At Greenwich, mounted in separate huts 50 feet north and south of main instrument. At Herstmonceux, in main pavilion 18 feet 6 inches north and south.

Pavilion: At Greenwich (in Christie enclosure, Fig. 45), semi-cylindrical in form, 20 feet wide by 30 feet long by 20 feet high. 8-foot aperture closed by two similar shutters. Outer skin of copper, inner skin of plywood and cork for thermal insulation.

At Herstmonceux, rectangular in form, 22 feet 6 inches wide by 50 feet long by 15 feet 6 inches high. The roof slopes 15° east and west, and carries two 50-foot shutters which open to 8 feet and shutters can be lowered to 9 feet 6 inches; 1 foot 3 inches above telescopic axis height. The construction of roof and walls is in duralumin sheeting on duralumin framework. The walls and roof are insulated with 4 inches thick slabs of ONAZOTE between the inner and outer skins.

Method of use

Generally similar to the procedure described on p. 45–6 with the following modifications; with photographic registration of circle-readings, only one observer is needed.

(1) *After step 1 (set telescope)*, set speed of RA micrometer drive to value of cosine declination.

(2) *Step 5 (RA observations)*
   (a) As star reaches RA wire, move two levers on eyepiece mounting, one to engage drive to start wire moving, the other to complete circuit for chronograph registration.
   (b) Immediately grasp control knob of differential gear and move travelling wire to bisect star image. Maintain bisection while micrometer is driven for about five revolutions: signal will be transmitted automatically to chronograph each revolution.
   (c) Return levers to normal position, disconnecting both drive and registration.

(3) *Step 7 (circle reading)*: Since 1952, this has been done photographically. As the button for operating the camera shutters is at the eye-end, the exposures can be made any time after the telescope has been clamped in Step 3. The films are removed for development after about 350 frames have been exposed.

49

(4) *Observations for measurement of level and zenith point* (nadir observations by mercury bath) are made at two-hourly intervals during each observing session. Collimation is measured once daily in the morning and also before and after each night session. Temperature readings of air and telescope and height of barometer are taken half-hourly during night sessions and more frequently during day observations, telescope is reversed once a fortnight.

(5) *Special instrumental observations*
  (a) Flexure of the telescope is measured at least once a fortnight via the collimators.
  (b) Three azimuth marks exist, one 3 miles south at Pevensey (PAM), another 1200 feet south (SAM) and the third 1400 feet north (NAM). All three marks are observed fortnightly, and in addition PAM is observed whenever azimuth stars are observed twelve hours later at the opposite culmination, day or night.

Historical Summary

1931: after 80 years of continuous service, Admiralty agreed to replace Airy Transit Circle; in the event, the latter remained in service for another 20 years. 1934: pavilion for new Reversible Transit Circle completed by Cleveland Bridge Co. in former Magnetic Observatory in Greenwich Park. March 1936: instrument erection completed; there had been considerable delays in design, particularly with weight-relieving gear. September 1936: transit observations started. August 1939: sapphire end-stop replaced by one of diamond. 13 September 1940: observations ceased because of enemy action; telescope lifted onto reversal pedestal where it remained until 1945. May 1945: telescope re-assembled. 1946: modifications to axis end-thrust and fixed eastern end-stop. 1948: modifications to eye-end. April 1952: new coma-free objective fitted. November 1952: RA circle-readings converted from visual to photographic. June 1953: dismantling at Greenwich and start of installation at Herstmonceux; the old pavilion at Greenwich was demolished and a new pavilion built at Herstmonceux. March 1954: pavilion building completed. 1955–56: alignment in new position. 28 November 1956: start of observations. 1957: installation of electronic punching and indicating chronograph (EPIC); produces machine-readable punched cards containing the five EPIC times of contact (RA) for each observation, serial number, observer's code, numerical name and culmination of object, position of ZD observation and special notes; the cards are processed by 10 a.m. next morning.

Contemporary Accounts

1973: K. C. Blackwell & M. E. Buontempo, *Second Greenwich Catalogue of Stars for 1950·0* (Royal Observatory, *Annals* No. 9, Herstmonceux, 1973). 1974: M. E. Buontempo, J. V. Cary & Patricia Eldridge, 'Provisional Positions of the Sun and Planets 1957–1971', *R. Obs. Bull.* No. 178.

5

The Large Altazimuth
Instruments

ALL the fundamental instruments so far discussed have been meridian instruments with which it is impossible to observe the Moon from four days before to four days after a new Moon due to glare from the Sun. As a consequence, there were parts of the Moon's orbit where she was never observed. Furthermore, it often happened that clouds obscured her at meridian passage but left her clear at other times.

Pointing out that the observatory had been founded expressly to study the Moon's motion, Airy in 1843 proposed a large Altitude and Azimuth instrument—an Altazimuth, essentially a transit circle that can be turned to any azimuth—so that accurate positions of the Moon could be obtained daily for comparison with the corresponding tabular places calculated from the lunar tables. That instrument and its successor are described here, portable altazimuths being dealt with in Chapter 9.

5.1.   *Airy's Altazimuth* (1847)

Designed by G. B. Airy.
Side castings inscribed: *Ransome and May, Engineers, Ipswich.*
Azimuth circle inscribed (at 360°): *Troughton & Simms, London.*
Mounted in New South Dome 1845–1910.
To Science Museum, London, 1929.
Displayed in Pond Gallery, Old Royal Observatory, from 1967.

Description (Figs. 47–49)

Airy's chief object in designing this instrument was to provide firmness by making it in as few pieces as possible, by using bolts not screws to join the principal parts, and by omitting adjusting screws, allowing the observations themselves to determine the errors which were then allowed for in computation.

The main components were as follows:

(a) *a 3-rayed brick pier*, 26 feet high, whose foundations were unconnected with the surrounding building; on the top of this was:

(b) *a cylindrical brick pier*, 3 feet in diameter, on which was mounted the horizontal (azimuth) graduated circle and the bottom pivot of the vertical axis, on which rotated:

(c) *the movable part of the instrument*, comprising four main iron castings—the two vertical cheeks and the upper and lower connecting plates. The cheeks carry (i) the micrometer-microscopes for reading the altitude and azimuth circles, (ii), the levels; and (iii), the lamp, reflectors and other apparatus for illuminating the telescope field and graduated circles; and (iv), the *Ys* for the horizontal axis of:

(d) *the vertical circles and telescope*, comprising the 3 feet graduated limb, 4 feet telescope and 3 feet axis.

The movable part of the instrument weighs just under a ton.

Method of use

An observation of one body comprises two readings of altitude, two of azimuth, one of each with the instrument facing east, one facing west.

1. Set telescope ahead of body to be observed. Assume azimuth is to be observed first and that instrument is facing east.
2. Switch on chronograph.
3. *1st azimuth observation*. As body passed obliquely across field, tap ivory key as body crosses each of 6 vertical wires. This will complete galvanic circuit and cause punctures on chronograph chart.
4. Read 4 micrometers and pointer micrometer on horizontal circle. Read levels.
5. *1st altitude observation*. Repeat 1 to 4 except that tap should be on 6 horizontal wires and vertical circle should be read.
6. *Reverse instrument* to face west.
7. Repeat 1 to 5.

Historical Summary

1843: order placed with Ransome & May of Ipswich for the engineering work and with Troughton & Simms of London for the optical and instrumental work. Some accident at Ransome & May delayed completion of the instrument. 1845: Altazimuth Dome built on the walls of Flamsteed's Observatory (at that time known as the Advanced Building). The tower was three storeys high and the dome itself cylindrical with sliding shutters (Fig. 48). 16 May 1847: first observation. 1854: chronograph brought into use. 1864: windows pierced in the south and east walls of the dome to improve ventilation. 29 November 1897: regular observations discontinued. 1899: new altazimuth brought into use.

1900–10: used periodically for observations of occultations of stars by the Moon, etc. 1910: dismounted; Dallmeyer photoheliograph mounted in the dome. 1939: to Science Museum, London, where it was exhibited until 1965. 1967: mounted in Pond Gallery, Old Royal Observatory.

Contemporary Accounts

1837: *The Illustrated London News*, 2 October 1847, 221–2. 1847: Airy, *Astron. Obs.* 1847, iv-xxvii. 1879: E. Dunkin, *The Midnight Sky* (1879), 158–61.

5.2.   *Christie's Altazimuth* (1896)

By Troughton & Simms.
Mounted in New Altazimuth Pavilion, 1896–1940.
No parts are known to have survived.

Description (Figs. 50–51)

This was virtually a reversible transit circle which could be placed either in the meridian or in any one of certain definite azimuths (say 45°, 60°, 70°, 80° or 90° E. or W.). It was then used essentially as a transit circle (except that stars, etc., crossed the field obliquely). giving altitude and azimuth directly for the same instant of time.* Unlike Airy's instrument there was no question of reversing the instrument during observations.

In addition to 15 horizontal and 15 vertical wires, the telescope had a position wire at the focus which could be set to the expected

*Christie, *Astron. Obs.*, 1908, xiv.

inclination of the path of the star across the field, but whose centre always passed through the centre of the field.

The OG was of 8 inches aperture and 8 feet focal length. Magnification 168 for stars, 180 for Sun (aperture reduced to 6 inches). There were two ZD circles of 3 feet diameter, one fitted on axis, the other movable. Two collimating telescopes were incorporated into the design of the dome. The instrument had its own chronograph. It was never a success.

Method of use

Remember that for observational purposes the telescope was movable in altitude but fixed in azimuth, and that the star (or any other body) crosses the field obliquely.

1. Before star enters field, set position wire to expected inclination of path across field.
2. When star enters field, move telescope in ZD to place star in position wire: clamp telescope.
3. Star should now run along position wire, passing through centre of field.
4. Observe times of transits across horizontal and vertical wires as star moves obliquely across field.

Historical Summary

1892: Christie proposed new altazimuth to replace Airy's 50-year-old instrument for extra-meridian observations. January 1896: New Altazimuth Pavilion and dome completed. To avoid interference with magnetic observations, site chosen was on the magnetic meridian of declination magnet and 90 feet north of it.* May 1896: Instrument erected under supervision of Mr Simms. Flexure of axis caused considerable delays. 23 February 1899: First regular observation. New chronograph used from March 1900.

1911: New mercury trough fitted, running on rails. 1923-7: Used in Prime Vertical to observe transits of fundamental stars. 1929: Last series of observations (Eros). 1940: Instrument dismantled and scrapped.

Contemporary Account

1908: Christie, *Astron. Obs.*, 1908.

*Christie, *Report*, 1892, 25.

6

The Zenith Instruments

ONE of the most fruitful sources of error in positional astronomy
is atmospheric refraction—the bending of light by the Earth's
atmosphere.

Primarily, the amount of bending varies according to the amount
of atmosphere the light ray has to pass through; a star 10° above the
horizon appears about 5′ too high, one at 50° appears only 50″ too
high. The amount of bending depends also upon the density of the
atmosphere (which varies with temperature and pressure) so the
effects of refraction can vary from day to day and hour to hour
according to the weather, the main cause of uncertainty.

But at the zenith, there is no bending. Where observations of
great precision are needed, therefore, zenith observations will
eliminate the uncertainties of refraction—a fact realized by
seventeenth-century astronomers trying to measure stellar parallax
to answer the question: "How far away are the stars?" In fact, even
the nearest star has proved to be much farther than was then thought,
and it was not until the nineteenth century that instruments became
available which were accurate enought to detect parallax.

Most of the instruments described in this chapter were designed so
that they always pointed vertically upwards with the centre of the
field of the telescope corresponding to the zenith, the only stars
which could be observed being those which pass within the field.
At Greenwich the principal star observed was the second-magnitude
γ Draconis which Airy called "the birth-star of modern astronomy"[1]
which, in Flamsteed's day, passed some 4′ north of the zenith at
Greenwich, and in Airy's day, some 2′ north.

However, in trying to detect parallax with his zenith sector, Bradley found two other previously undetected periodical variations in the positions of stars—the aberration of light due to the velocity of the Earth in its orbit relative to the velocity of light, and the nutation (or nodding) of the Earth's axis due to the varying effects of the Moon upon the forces which cause the precession of the equinoxes.

Later zenith instruments were used to check the line of collimation of the meridian telescope, to measure the constants of aberration and nutation, the effects of polar wander, and, since the 1950s, for time-determination.

6.1. *Flamsteed's Well Telescope* (1676)

Object-glass by Pierre Borel of Paris.

Only recorded observations 1679.

The object-glass is preserved in the Science Museum, London, inscribed: *Flamstead's O.G. 90ft fo[s] presented by Jm[s] Hodgson Esq Nov 3 1737 Royal Society No 25.*

Description (Figs. 52 and 53)

In principle the design of this telescope was both ingenious and simple: the long-focus object-glass in its cell was set into the stone-work at the top of the 100-foot well pointing vertically upwards; suspended from it was a plumb-line with the telescope eyepiece (with presumably some sort of micrometer) acting as plumb-bob; the plumb-line was enclosed in a wooden tube; the observer reclined on a couch at the bottom of the 'well', looking upwards through the plumb-bob eyepiece.

Though details are scanty, there are a few firm facts. The telescope was intended for finding the parallax of γ Draconis which in Flamsteed's time passed within about 4' of the zenith at Greenwich. (We now know that its parallax was of the order $0''\cdot017$ so it is not surprising that no firm results were obtained.) Flamsteed and Moore hoped to be able to use it in day time as Robert Hooke had done with his Gresham College zenith telescope.[2] Several different recesses had to be prepared for the OG in the stonework of the cupola to allow for the uncertainty of the focal length.[3] Sir Jonas Moore obtained the OG from Pierre Borel (1620–89) the French Academician, physician and chemist.[4] Because the OG has survived (it was presented to the Royal Society in 1737 and is now on loan to the Science Museum, London) we know that it was of very poor quality, was

9·7 inches in diameter, 0·36 inches thick, weighed 39½ oz. and had a focal length of about 87 feet 5 inches.[5]

However, probably because the experiment did not prove to be a success, many questions remain unanswered. When was the 'well' dug? When was it filled in? What was the form of the eyepiece, micrometer and suspension? Why was the experiment abandoned?

Because there is no documentary evidence about the digging of the shaft in Flamsteed's time, it seems likely it was originally connected with Greenwich Castle.

As to its position, a filled-in shaft was discovered about 1790 when the wall to the middle garden was being built. In 1965 excavations revealed a filled-in shaft of the correct diameter (about 7 feet) exactly in the position marked by Flamsteed himself in the plan in RGO MSS. 18/3. The foundation of the wall above had been arched over, presumably in 1790. However, in the 25-foot depth which was excavated in 1965, no evidence of a brick lining was found.[6]

When James Hodgson, F.R.S. (1672–1755)—who was Flamsteed's assistant from 1695 to 1702, married his niece and became one of his executors—presented the OG to the Royal Society in 1737, the experiment was said to have failed "because of the damp of the place".[7] The difficulties of bringing a 90-foot plumb-line to rest, the poor quality of the OG, and Sir Jonas Moore's death in August 1679 —all these factors probably contributed to its abandonment.

We have almost no details of the plumb-line suspension or the plumb-bob/eyepiece beyond what can be seen in Fig. 52 and some scanty references in letters to Sir Jonas. Flamsteed used an 8-inch focus eye-glass (giving a magnification of about 130) for finding the focal length before the OG was mounted.[8] As it would have been impossible to touch the plumb-bob/eyepiece when observing, the micrometer was presumably some form of reticle. This seems to be confirmed by Flamsteed in his 1678 letter to Sir Jonas: "I am glad my contrivance for the eye-glass and thrids so satisfies you".[9]

Historical Summary

Probably 1676: well telescope illustrated in Francis Place's series of etchings (see Figs. 5 and 52). In the *Ichnographia* (plan) the well (marked *Puteus profunditatis 120 ped: cum Tubo pro Observ: Parallaxis Terrae,*) is shown in the wrong position. November 1677: Flamsteed measuring focal length and arranging alternative positions for OG.[10] 17 February 1678: Moore discusses well telescope with Hooke.[11]

7 March 1678: Flamsteed promises observations will start as soon as
γ Draconis passes zenith in the dark hours.[12] 20 February 1679:
experiment with barometer to note changes in height between roof
of Great Room and bottom of well.[13] 20 June and 30 July 1679:
observations recorded. August 1679, death of Sir Jonas Moore.

3 November 1737: OG presented by James Hodgson to Royal
Society. About 1790: in about 1790 when they were digging the
foundations for a wall, which formed the south boundary of the
garden, the men came to a well, but unfortunately built over it
immediately before anyone could see it.

1881: a subsidence revealed further evidence of brick work; no
action taken.

1932: OG lent to Science Museum, London. 1965: excavation to
25 feet by the Lewisham Natural History Society under superintendence of Ministry of Public Building & Works.

6.2. Bradley's 12½-foot Zenith Sector (1727)

*The instrument with which James Bradley discovered the aberration
of light and the nutation of the Earth's axis*

By George Graham of London.

Mounted at Wanstead 1727–49.

Mounted at Greenwich 1749–1837.

Used at Cape of Good Hope 1837–39.

Displayed in ORO since 1960. Now in Quadrant Room.

Description (Figs. 54–57)

As outlined in the Historical Summary below, this instrument was
specifically designed to investigate the unexpected results obtained
by Bradley and Molyneux in 1725 in observations to try to detect
parallax in γ Draconis, made at Kew in the latter's 24¼-foot zenith
telescope—discordances which were later found to be due to aberration and nutation.

Its original purpose was to detect any small changes in the ZD of
stars near the zenith, which might occur from day to day or year to
year. Later, because its potential accuracy was about one-tenth of a
second of arc as compared with Maskelyne's claim of 10″ for the
Greenwich mural quadrants, it was used for settling the errors of
collimation of the telescopes of the latter.

The main essentials are shown in Fig. 54. The 12½-foot telescope is
suspended vertically from a horizontal east–west axis *a* so that it can

be moved pendulum-like in the meridian up to $6\frac{1}{2}°$ from the zenith —a limit chosen to permit observations of the bright star Capella.

Fixed to the wall near the eye-end is the back-arch *h* on which is clamped the slider carrying the micrometer. The micrometer screw *f* is kept bearing against the telescope tube by means of a weight fastened to a cord passing over a pulley at *k*.

A plumb-line, suspended from the centre of the axis *a*, passes in front of the $12\frac{1}{2}°$ graduated arc attached to (and therefore moving with) the telescope. As the arc is graduated only at 5′ intervals, each ZD measurement has to be made in two steps, as explained below.

The bracket *ik* supported the wooden plumb-line guard (made with doors like a cupboard—it has not survived) and the pulley for the counterpoise weight.

Details are as follows:

Telescope: tube of iron weighing 8 lb 15 oz. OG $3\frac{1}{2}$-inch aperture, 12 feet $6\frac{1}{2}$ inches focal length until achromatic lens fitted 1801, then 12 feet $9\frac{1}{2}$ inches. Magnification 70.

Limb: orginally, brass arc graduated every 5′ in NPD as at Wanstead with limb facing west. In 1781, 'in order to avoid the correction in the Parallactic Sector necessary on account of the different dilations of the Iron tube and the brass arch; and also on account of the inaccuracy of the present divisions', Jeremiah Sisson replaced Graham's brass arc by one of steel, with divisions every 5′ of ZD upon gold pins.

Micrometer: graduation a fraction less than 1″ apart; one revolution equals 33″·6.

Plumb-line: before 1765, silver wire 1/100 inch thick; from 1774, gilt-silver wire 1/218 inch thick. Plummet 18 oz. In 1768, Bird "altered the Manner of Suspension of its Plumb-line, which before hung from a Notch made precisely at the Centre of the Instrument, but is now suspended from a Notch a little above the Centre; and the Notch is moved by means of a Screw, so that the Silver Wire, which is the Plumb-line, may appear through a Microscope to pass over and bisect a fine Point placed at the Centre".[14]

Instrument supports: its later rôle at Greenwich demanded it be reversed periodically. Two supports were needed, one facing east, one facing west. However, because of the hip roof, the west wall of the Quadrant Room was not high enough to take the Sector, so, when reversal was needed, it had to be taken down from the

Quadrant Room, carried through two doors and along a passage, and re-erected in the Transit Room—an operation which took the whole morning and was not without hazard. The openings are marked *S* and *s* in Fig. 6.

In 1779, the west wall of the Quadrant Room was raised to allow a support to be put on it. The sector was thereafter used only in the Quadrant Room. When used at the Cape of Good Hope 1837–9, the instrument was mounted on a 'great tripod' on which it could be reversed.

Method of use (Figs. 54 and 56)
Taking an observation
1. Set telescope to approximate expected polar distance of star to be observed after unclamping slider from back arc. Re-clamp slider.
2. Make plumb-line bisect nearest gold dot on sector arc by turning micrometer screw.
3. Record which point on sector arc is bisected by plumb-line in column (*A*) of Fig. 56: record micrometer reading in column (*B*).
4. Look through telescope and wait for star to come into field of view.
5. Bring star onto horizontal (east–west) crosswire by moving micrometer screw.
6. Record micrometer reading in column (*C*) at moment of transit, i.e., when star reaches the vertical (north–south) crosswire.
7. Make plumb-line bisect nearest gold dot once again, by moving micrometer screw.
8. Record micrometer reading in column (*D*). This is merely a check on reading in column (*B*).

Calculating the North Polar Distance
9. Find difference of micrometer readings between the star in column (*C*) and the gold dot (a mean of columns (*B*) & (*D*)); enter in column (*E*).
10. Convert micrometer difference in column (*E*) into minutes and seconds of arc, using a conversion table, and enter in column (*F*).
11. Required NPD is the sum of (*A*) and (*F*).

Historical Summary

As we have seen, seventeenth-century efforts to detect parallax were not entirely successful. In 1725, therefore, Samuel Molyneux (1689–1728) set up at Kew a 24¼-foot zenith sector, made by Graham specifically for observing γ Draconis, in order to try whether it had any feasible parallax.

Erected in November 1725, the first observation was obtained on 3 December by Molyneux and Bradley working together. Only 14 days later, Bradley noticed that the star seemed to be moving slowly southwards from day to day, whereas parallax should have caused it to move northwards. During the year 1726, this phenomenon—which we now know to have been due to aberration—was confirmed by the two astronomers though the cause was not then known.

With Molyneux's sector, Bradley could only investigate stars which passed within 7′ or 8′ of the zenith so, early in 1727, he decided a more versatile instrument was needed: Graham set about designing a new sector, similar in principle to Molyneux's but capable of observing stars up to 6¼° from the zenith, thereby including the first magnitude Capella.

The new telescope was erected on 19 August 1727 in the house of Bradley's aunt, widow of astronomer James Pound, in Wanstead, Essex. Observations with it soon confirmed previous findings.

Supposedly inspired by the behaviour of the masthead pennant when sailing in a boat on the Thames in September 1727,[19] Bradley described the causes of aberration in a paper he read to the Royal Society in January 1729, just three years after he had first noticed its effects.

He continued to make observations with the sector at Wanstead until 1747, in which year—having observed γ Draconis and other zenith stars during one complete revolution of the Moon's nodes—he was able to announce his second great discovery, the nutation of the Earth's axis.

Subsequent history of the sector is as follows: 1742: Bradley became Astronomer Royal at Greenwich. 1749: sector purchased by the Government for £45 and moved from Wanstead to Greenwich; alternative hanging positions (*S* and *s* in Fig. 6) in Transit and Quadrant Rooms, put up by Hearne who had made the brackets in Wanstead in 1727; at this time, it was often called a 'parallactick sector'. 1768: plumb-line suspension modified by Bird. 1779:

alternative support moved from west wall of Transit Room to west wall of Quadrant Room, which had been specially raised. 1781: steel arc with gold points by Jeremiah Sisson replaced the old brass arc by George Graham. 1785: plumb-line guard and plummet pot fitted; new crosswire-illumination arrangements.

1801: achromatic object-glass fitted by Peter Dollond—£11 11s. 16 December 1812: last recorded observation at Greenwich. 1837: sent to the Cape of Good Hope for verification of the length of an arc of meridian measured by Lacaille in 1750. 1839: returned to Greenwich. 1851: hung as relic on west wall of Transit Circle Room (Fig. 73).

1967: moved to 1779 position on west wall of Quadrant Room.

Contemporary Accounts

1729: J. Bradley 'A letter to Dr. Edmund Halley, Astronomer Reg. & C giving an account of a new-discovered Motion of the Fixed Stars'. *Phil. Trans.*, XXXV (1729), 637. 1776: N. Maskelyne, *Astron. Obs.* (1776) Introduction (ix)-(xi). 1790: S. Vince, *A Treatise on Practical Astronomy* (1790) 148–51.

1832: S. P. Rigaud, *Miscellaneous Works and Correspondence of James Bradley* (Oxford, 1832), 194–200, 201–2. 1866: Sir T. Maclear, *Verification and Extension of La Caille's Arc of Meridian* ... (1866) (description and history by Airy).

6.3. *Pond's 9½-foot Zenith Tube (or Zenith Micrometer)* (1812)
By Edward Troughton, London.
Mounted in Circle Room 1812–16.
Exhibited in Bradley's Transit Room since 1967.

Description (Fig. 28)

Designed to determine the zenith point of the mural circle by comparing observations of γ Draconis, this was a 6-inch Newtonian reflecting telescope fixed in the zenith with a plumb-line passing through the centre of the tube. In the eyepiece, it was theoretically possible to see both the plumb-line and the star to be observed (providing it was within 15 arc-minutes of the zenith). Its zenith distance could then be measured by micrometer. When one reading had been taken, the telescope was turned through 180° and another reading of the same star taken. In fact, the illumination of the plumb-line and micrometer was so defective that the telescope was never used successfully.

Today, the plumb-bob and brackets are modern, the optics are missing.

Historical Summary

1812: telescope fixed on the back of the mural circle pier as in Fig. 28; it was made to appease those members of the Royal Society who criticized the mural circle for having no built-in vertical reference such as a plumb-line; never used successfully because the illumination was defective. 1816: replaced by 8-foot achromatic zenith telescope. 1851: mounted horizontally on west wall of Transit Circle Room as a relic.

1960: exhibited in Herschel Gallery, Flamsteed House. 1967: mounted in present position.

Contemporary Account

1829: W. Pearson, *An Introduction to Practical Astronomy*, II (1829) 473, 479 and 480.

6.4. *Pond's 8-foot Achromatic Zenith Telescope* (1816)
By P. & J. Dollond, London.
Mounted in Circle Room, 1816–46.
Exhibited in Bradley Transit Room since 1967.

Purpose

To determine the zenith point for the mural circle in place of the defective Newtonian telescope mentioned above. The need for it disappeared in 1822 when it was realized that the zenith point could be determined more accurately by reflection observations with the mural circle itself.

Description

Object-glass: achromatic, by Peter Dollond, 7 feet 10 inches focal length, 2¾-inch aperture, taken from the 8-foot transit instrument in which it had been mounted from 1774 until 1816.

Eyepiece and micrometer now missing. No recorded observations have been found, so its method of use is uncertain.

Historical Summary

1816: replaced 9½-foot Newtonian zenith tube (whose design had proved defective) on back of circle pier; the Alpha Aquilae telescope (see p. 117) was presumably mounted at the same time. 1833: superseded by 25-foot great zenith sector mounted in Flamsteed House; the 8-foot telescope remained on the Circle pier. 1846: dismounted. 1851: hung horizontally as relic on west wall of Transit Circle Room.

1960: exhibited in Herschel Gallery, Flamsteed House. 1967: mounted in Bradley's Transit Room on replica pier; when in use it was a few feet east of there.

Contemporary Accounts

Phil. Trans., 1817, 158, 353; 1818, 477. Pond; *Astron. Obs.*, II, 371–385.

6.5. *Pond's 25-foot Great Zenith Tube* (1833)

By Troughton & Simms of London, OG by Dollond.

Mounted in Zenith Sector Apartment at N.W. corner of Flamsteed House, 1833–48.

Part of OG survives.

Description (see Fig. 58)

Variously called the Great Zenith Sector, Zenith Tube, or Zenith Micrometer, this instrument was primarily designed to define the Zenith point for the mural circles by observations of γ Draconis.

Basically, it consisted of a refracting telescope of 24 feet 6 inches focal length, 5-inch aperture, mounted in the zenith with the vertical defined by a plumb-line. The 1840 inventory describes the way of mounting.

> The Zenith-sector of 25 feet, resting by its lower point upon a conical stone capped with two circular brass plates which are adjusted with 4 screws, and supported at its upper part by the top of a cast-iron tube (which incloses the telescope and stands upon four arched legs).[20]

A filar micrometer at the eye-end measured the ZD of γ Draconis at culmination. The telescope could be turned through 180° but this operation took not less than 15 minutes (as originally designed, the instrument had to be placed precisely in the vertical each observation) so that there was never any question of taking a pair of observations at the same culmination.

The instrument was never a succeess. Unevenly heated air over the OG caused troubles which were not cured, despite modifications to the building in 1834, 1837 and 1844–5. The weight of the original micrometer and eyepiece caused flexure of the telescope which was cured by Airy in 1837. At the same time, he modified the plumb-line assembly so that the telescope did not have to be placed precisely vertical before each observation: micrometer-microscopes were fitted so that the departure from the vertical could be measured and allowed for in computation.[21]

Historical Summary

July 1819: Board of Visitors recommended Edward Troughton provide new sector. 1820: Zenith Sector room erected at north-east corner of Flamsteed House, on site of Halley's transit room, long since demolished; order placed with Troughton (who took William Simms into partnership in 1826). 1820–31: considerable delays; up to 1822, these seem to have been caused by disputes over advance payment. June 1831: Great zenith sector erected; first recorded observation 1 August, clock, 'Arnold 2.' May–July 1834: new roof placed on building to avoid convection currents. 1837: Airy made considerable modifications (see above). 1 May 1848: last observation; instrument dismounted; room incorporated into dwelling house. 1850: OG let into north wall of Airy's altazimuth dome for use as collimator. Partially destroyed by AA shell splinter in World War II. Remains now preserved in NMM.

Contemporary Accounts

No full account has ever been published. The description here has been inferred from the references given and from the published observations.

6.6. *Airy's Reflex Zenith Tube* (1851)

Designed by G. B. Airy. Made by William Simms. OG (1793) by Peter Dollond, ex 10-foot transit instrument.

Signed: *Troughton & Simms, London, 1851.*

Mounted in own room near transit circle, 1851–1923.

Displayed in Pond Gallery, Old Royal Observatory, since 1967.

Description (Figs. 60–62)

Designed by Airy for observations of γ Draconis, this instrument used his double-zenith-distance method, entailing two rapid observations as it crossed the meridian, the OG and micrometer being turned through 180° between observations.

The optical principles are shown, somewhat simplified, in Fig. 60. A pencil of light from γ Draconis when overhead passes through the OG *AB* and is reflected from a mercury surface *C* placed beneath at a distance somewhat less than half the focal length. Passing through the OG a second time, it comes to a focus just above the OG, at the plane of the micrometer wires.

Turning to Fig. 62, micrometers A and B are both able to move the same 30 wires across the field. The eyepiece (5) is focused on the micrometer wires via the prism (8) (one component of the eyepiece is actually fixed beneath the prism).

It so happens that micrometer wires 15 and 16 were placed at an interval corresponding to the double-ZD of γ Draconis. Latterly, the other 28 wires—actually spiders' webs—were omitted altogether.

The rotating head (4) carrying the OG (6) and micrometer (7) could be rotated 180° in azimuth so that the micrometer A was either to the left or right of the observer (who sat facing west), the eyepiece (5) and prism (8) remaining fixed.

Focusing was done by varying the distance between the OG and the mercury surface.

The achromatic OG (originally made by Peter Dollond in 1793 and used in Troughton's 10-foot transit from 1816 to 1850) had a focal length of 9 feet 8 inches and an aperture of 5 inches.

Method of use (Fig. 61)

Assuming Micrometer A was to the observer's right hand at the start and that star was γ Draconis.

1. Read and record Micrometer B before star enters field.
2. When star appears in telescope field, place Wire 15 on star image by means of Micrometer A.
3. Without waiting to read Micrometer A, reverse the instrument head and then place Wire 16 on star image by means of Micrometer B.
4. When star has passed away, read and record both micrometers. Read spirit level.

Reduction is made by the correct addition or subtraction of the micrometer readings to the known distance apart of Wires 15 and 16

i.e. *1st Observation:* reading of A+first reading of B.

2nd Observation: reading of A+second reading of B+ interval between Wires 15 and 16.[18]

A further correction is applied for any inclination of the OG out of the horizontal as indicated by the spirit level.

Historical Summary

1850: Reflex Zenith Tube room prepared as annexe to Transit Circle Room, off its north-east corner. 1851: instrument mounted; first recorded observation 9 September. 1856: because of vibration on the surface of the mercury, a new room was built off the south-west corner of the Transit Circle Room; clock, 'Mudge & Dutton'. 1899: RZT observations ceased due to pressure of work.

1902: new series of observations started to determine the amount of latitude variation, something which had been discovered from analysis of observations on this instrument from 1882–86. 1911: last regular observation 24 August; superseded by Cookson Floating Zenith Telescope (lent by Cambridge Observatory). 1923: RZT dismounted. 1937: transferred to Science Museum, London. 1967: mounted in present position in Pond Gallery, Old Royal Observatory.

Contemporary Accounts

1854: Airy, *Astron. Obs.*, 1854, Appendix I. 1854: Inventory 3 June 1854 (RGO MSS.), f.17.

6.7. *Airy's Water Telescope* (1870)

Designed by G. B. Airy.

Signed: *Troughton & Simms London.*

Mounted on South Ground, 1871–73.

Displayed in Old Royal Observatory since 1960.

Purpose: To measure the coefficient of aberration when γ Draconis was observed through approximately 36 inches of water.

Method: Double zenith distance observations similar to those performed in the RZT.

Description: Light collected by specially corrected OG is passed through water-filled tube and focused upon the wires of the double-headed micrometer, and viewed by rotating elbow eyepiece. Telescope is supported on a hollow spindle which is supported at either end and allowed to rotate in a rigid frame. Provision made for locating the telescope on its spindle in either of two positions 180° apart by rotating spindle until screw or its mate on opposite side of spindle is against stop face. Error from vertical in either position checked by spirit levels. The whole unit is mounted upon a stone pillar.

Method of use

The procedure is very similar to that described above (pp. 68–9) for the RZT.

Historical Summary

Professor Klinkerfues in 1867 had performed an experiment with an 8-inch tube of turpentine inserted into the optical path of a transit instrument. His findings led Airy to believe that the coefficient of aberration might be sensibly altered when light passed through any other refracting medium than air; e.g. the thickness of the OG in Airy's transit circle might have sufficient effect to be measurable.

Airy therefore designed a special zenith telescope, whose tube could be filled with water, to carry out an experiment to prove or disprove this theory. Made by Troughton & Simms, it was mounted in Struve's Occasional Observatory on the South Ground in June 1870.

Observations between February 1871 and July 1873 proved that the amount of sidereal aberration did not depend upon the medium through which the light passed—a result which was of great significance in the undulatory theory of light. The telescope was dismounted in August 1873. It can now (1975) be seen in the Old Royal Observatory, Pond Gallery.

Contemporary Description

Airy, G. B., 'The History and Description of the Water Telescope', *Astron. Obs.* 1871, Appendix.

6.8. *Cookson's Floating Zenith Telescope* (1900)
Designed by Bryan Cookson, M.A., of Cambridge.
Signed: *The Scientific Instrument Co Ltd Cambridge.*
OG by Messrs Cooke & Sons of York.
Trough and float by Norman Cookson and Hon. C. A. Parsons.
Mounted at Cambridge, 1900–11.
Mounted at Greenwich, 1911–51.
At Science Museum, London, 1956–65.
Displayed in Frank Dyson Gallery, Old Royal Observatory since 1967.

Purpose

To determine aberration and nutation constants and to measure latitude variation ('Chandler wobble'—the wandering of the Earth's poles, up to 30 feet from the mean position) by photographing selected pairs of stars that transit at approximately equal distances north and south of the zenith—the Talcott method.

Description (Fig. 63)

There are three principal parts—the trough (1), the float (2) and telescope (3).

The trough and float, both of cast iron, are shallow annular basins; the float is concentric with the trough, and fits inside it with half an inch clearance all round. Mercury (142 lb) is poured into the trough, and the float is free to take up its position of equilibrium in the bath of mercury.

The telescope (3) passes through the centre of the annulus, and its trunnions (4) rest in Vs which are carried by the float (2). The trough (1) is fixed, but the float (2) is free, and can be rotated in azimuth. Two knife-edge indexing devices (5), capable of fine screw adjustment, ensure that the telescope is exactly in the meridian—either north or south—when exposures are made.

The telescope tube is set to the required ZD by means of the clamp (6), sector (7) and setting circle (8). The plate holder (9) locates the 5 inch × 2 inch plate in the focal plane of the OG, which is a Cooke triplet of 6-inch aperture and 65·4-inch focal length. At Greenwich, exposures were made by a sliding shutter on the roof.

Method of use

Pairs of stars between declination +30° and +70° with photographic magnitudes between +4½ and +8 were chosen.

Each plate would first photograph the trail of a star across the meridian to one side of zenith, then the telescope would be swung round to photograph the paired star in a similar fashion. The pairs of stars chosen would be so arranged in RA that some minutes would elapse between meridian transits of each star of the pair to allow ample time to rotate the instrument.

A similar plate at a future date would be taken and compared with the first by a measuring micrometer. A change in latitude (zenith point) or a change in declination due to aberration or nutation shows as a displacement between the pairs of star trails. The meridian shows on the plate as a break in the star trails caused by 0.1 mm wire fixed in the meridian at the focal plane.

Historical Summary

1900: mounted in 15-foot dome on top of main building in Cambridge University Observatory; used for two years by designer, Bryan Cookson, to investigate aberration constant. 1911: lent to RO by Cambridge Observatory; mounted in own hut in N.W. corner of courtyard; observations for latitude variation and aberration constant. June 1936: moved to new house in Christie enclosure near RTC; new programme for latitude variation only. September 1940: observations discontinued due to war conditions; minor damage in air raid in October when exposing shutter in roof fell on telescope. 1951: dismantled. 1956: to Science Museum, London (Inv. No. 1956–62). 1967: to NMM, Dyson Gallery, on display (NMM Acq. No. NA65-42L).

Contemporary Accounts

1901: Bryan Cookson, 'Description of a Floating Photographic Zenith Telescope and some preliminary results obtained with it', *MNRAS* LXI, 5, (March 1901), 315–21. 1939: H. Spencer Jones (Ed.), *Observations made with the Cookson Floating Zenith Telescope in the years 1927–1936* . . . (Greenwich 1939).

6.9. *Photographic Zenith Tube* (1955)

By Sir Howard Grubb, Parsons & Co. Newcastle-on-Tyne, designed by D. S. Perfect.

Mounted at Herstmonceux, 1954.

Still in use, 1975.

The Zenith Instruments

Purpose

As a transit instrument for time-determination and measurement of latitude variation. Its use is limited to stars passing within 15′ of the zenith but this drawback is offset by the superior accuracy of the measurements. At the beginning of the century, accuracy expected in time-determination with a conventional transit instrument was about one tenth of a second of time (100 milliseconds): the PZT measures to 20 milliseconds for a single star, 4 milliseconds for a night's work of, say, 30 stars.

Description (Figs. 64 and 67)

The essential principle of this instrument is the same as that of Airy's Reflex Zenith Tube described on pp. 67–9 above, except that the eyepiece is replaced by a photographic plate: the converging light from a star very close to the zenith, having passed through the OG L, is reflected at a mercury surface M and brought to a focus on a photographic plate P placed just below the OG. The instrument is thus insensitive to errors of level and, being 'folded up' by the reflection principle, can be made far more rigid than a conventional transit instrument of the same focal length.

Each observation consists of four exposures of a single star, each of 20^s—two before meridian transit, two after. Turning the plate 180° between each exposure allows the zenith point (and thus the latitude) to be found in a manner similar to that used with the Cookson Floating Zenith Telescope (pp. 71–2), though plates from the latter instrument showed trails from two different stars some way from the zenith whereas those from the PZT show near-instantaneous exposures of a single star very close to the zenith. The times of each exposure are accurately (and automatically) recorded by chronograph.

Thus the final picture on the plate is a nearly rectangular array of four dots, as illustrated in Fig. 66. Broadly speaking, measurement of the plate in x (E–W) direction combined with chronograph times yield time of transit; measurement in the y (north-south) direction yields latitude.

Object-glass: 25 cm aperture, 347 cm focal length, doublet flint-ahead, N2 point 1·2 cm behind rear component.

Mercury pool: 170 cm below lens, to reflect focus to N2 plane.

Plates: 41 × 41 mm, glass; exposure 20^s.

73

Method of use

The operator—no longer an observer—sits at a console in a separate building (Fig. 68). For each observation the operation of the telescope is entirely automatic.

As the stars differ in ZD, it is often possible to use the same plate for several observations—say 20 images of 5 stars.

Historical Summary

1944: greatly increased accuracy in timekeeping achieved by quartz crystal clock demanded far higher accuracy in time-determination than the small reversible transit was capable of; order placed with Grubb, Parsons for PZT. 1952: instrument assembled at Newcastle. November 1954: installation commenced at Herstmonceux. Autumn 1955: observations commenced. 16 October 1957: PZT superseded small transit C for time-determination.

Contemporary Account

1958: Perfect, D. S., 'The Photographic Zenith Tube at the Royal Greenwich Observatory', *Occ. Notes R. Astron. Soc., 21* (November 1959), 223–33.

NOTES

1. G. B. Airy, "History and Description of the Water Telescope. . . .", *Astron. Obs.*, 1871, Appendix 1.
2. R. Hooke, *An Attempt to Prove the Motion of the Earth* . . . (1674), pp. 11–25.
3. RGO MS. 36/48, Flamsteed to Moore 14 November 1677.
4. *Phil. Trans.*, XI, (1677), 691.
5. A. Hunter and E. G. Martin, "The Flamsteed 90-foot lens", *The Observatory*, 76 (February 1956), 24–6.
6. Lewisham Natural History Society. *Darenthis*, IV (1966), pp. 7–17.
7. RS C.M., 3 November 1737, p. 129.
8. RGO MS. 36/48, Flamsteed to Moore 14 November 1677.
9. *Ibid.*
10. *Ibid.*
11. Hooke, 17 February 1678.
12. RGO MS. 36/55, Flamsteed to Moore, 7 March 1678.
13. RGO MS. 43/50.
14. RGO MS., letter Margaret Maskelyne to Mrs Pond.
15. Maskelyne, *Astron. Obs.* I (quadrant), 38.
16. RS MS. 651/9, 1840 Inventory, Zenith Sector Apartment No. 1.
17. Airy, *Astron. Obs.*, 1837.
18. Airy, *Astron. Obs.*, 1854, Appendix I, viii.

7

The Equatorial Sectors

Two somewhat different types of instrument are described in this chapter. Before the days of the pendulum clock, the most accurate means of obtaining differences of RA was by large-radius sector such as the sextants of Tycho and Hevelius. Flamsteed used his sextant for taking 'intermutual distances' (his own words), primarily for compiling his star catalogue. He abandoned it in favour of a mural arc and pendulum clock in 1689.

The eighteenth-century instruments described here—sometimes called astronomical sectors—were different in principle, being designed specifically for finding the positions of comets by comparison with a nearby star whose RA and Dec were known. They were superseded when large telescopes—such as the Shuckburgh and Sheepshanks at Greenwich—came to be mounted equatorially and fitted with accurate circles of RA and Dec.

7.1. *Flamsteed's 7-foot Equatorial Sextant (1676)*

Designed largely by John Flamsteed. Framework supervised by Edward Sylvester of Tower of London; wheel-work, limb and indices by Thomas Tompion of London; divisions by Flamsteed.

Mounted in Sextant House 1676–1720.

No record after 1720.

Description (Figs. 68–70)

Flamsteed claimed his sextant was superior to the equivalent instruments of Tycho and Hevelius in having the following features:

(*a*) *A polar instead of a vertical axis:* once the left-hand object was found in the fixed telescope, it could be kept in the field simply by pushing on the telescope to follow the Earth's rotation, regardless of any changes made in the inclination of the sextant to pick up the right-hand object in the moving telescope: once both objects were acquired, the intermutual distance could be measured at leisure.

(*b*) *Telescopic instead of open sights:* not only did this make for great accuracy of measurement but it permitted planets and bright stars to be observed in daytime: Flamsteed could measure the distance between Venus and the Sun when the former was only 16° from the Sun: Hevelius's minimum with open sights was 40°.[1]

(*c*) *Sextant attached to axis by toothed semicircles instead of ball-and-socket joint:* the setting of the greater semicircle (graduated by Flamsteed in 1677) governed the declination of the fixed telescope, that of the lesser semicircle (not graduated so far as we know) the inclination between the lines of sight of the two telescopes.

(*d*) *Perpetual screw instead of clamp and tangent screw for moving telescope:* a worm working on an endless screw on the edge of the limb (but capable of being withdrawn for initial setting), exactly as in the modern 'micrometer' marine sextant, instead of clamp and tangent screw as in today's 'vernier' sextant: Flamsteed's original design envisaged the earlier method but Hooke persuaded Moore that the perpetual screw (which Hooke claimed to have invented—falsely according to Flamsteed) should be adopted.

(*e*) *Graduations by 'revolves' of the screw instead of degrees and parts of a circle:* originally the measured angle was indicted by the number of the revolves and parts of a revolve of the screw (the threads on the edge being numbered on the face of the limb); conversion to degrees and parts of a circle being by table[2]: in December 1677, Flamsteed personally divided the limb into degrees and minutes with a diagonal scale: thereafter he always read both scales.

(*f*) *Ample space for both observers' heads when measuring small angles.*

Fig. 68 illustrated Flamsteed's own description of the instrument[3] where he described the parts as follows:

A   round plate 9 in. wide, 1¼ in. thick.
BB   fixed telescope.
d   seven small struts of solid iron.
e   ten spokes, 1 in. wide and 1½ in. thick where they are attached to plate *A*, narrowing towards their outer ends where they are fixed.

FF   greater (declination) semicircle, toothed on outer edge, attached to *NN*.

G   lesser semicircle.

H   iron bar on frame parallel to *BB*: *NN* is attached to it by *MM*.

IK   movable index and telescope.

L   limb made of iron $1\frac{1}{2}$ in. wide, covered with brass plate of the same width and $\frac{1}{4}$ in. thick; the outer edge is of screw-threaded steel, 17 threads to the inch.

11   bar perpendicular to *NN*, supporting the lesser semicircle *G*.

M   two thick bearings of brass attached to *H* which receive the two ends of the diameter *NN*.

m   observing steps.

NN   diameter to which *FF* is fixed.

o   Octagon Room.

PP   polar axis of wrought iron, diameter about 3 in., lower part round, top almost square, with opening for declination semicircle *FF*; vertex is at centre of *FF*.

Q   half-shaft carrying *PP* on top of *S*.

R   pedestal carrying bottom of polar axis *PP*.

S   block of solid oak supporting *Q* and *PP*.

xy   worm gear and handle working declination semicircle *FF*.

γ   iron fastening attached to *FF* through which passes the lesser semicircle *G*.

Method of use

In its normal rôle—that of measuring intermutual distances—the sextant required at least three persons to manage it, the senior observer on the moving telescope observing the right-hand object, the second observer on the fixed telescope, observing the left-hand object and, for a third, 'any indifferent person, of a strong able body',[4] to manipulate the controls of the two semicircles.

Whenever possible, measurements were made when the two bodies were at similar altitudes so as to avoid the complication of refraction affecting the two bodies differently. However, this could not always be done and, in the 1684 example used for Fig. 69, Venus was at 26° altitude, Pollux at 63°.

Because Hooke's mural quadrant proved unworkable, Flamsteed sometimes used his sextant after 1677 in a rôle for which it was not designed. Suspending a plumb-line from the centre, he set the sextant precisely in the meridian and used it to measure meridian zenith distances.[5]

Making an observation (see Fig. 69)

This is an example of an actual measurement of the distance

between Venus and Pollux, taken by Flamsteed and probably Abraham Sharp, at 6 a.m. on 21 September 1684.

The 2nd observer:
(1) set greater semicircle to declination of left-hand object (Venus);
(2) pointed fixed telescope at Venus by turning sextant about its polar axis (to the appropriate hour angle) by pressure on frame of sextant;
(3) looked through fixed telescope and brought image of Venus on to telescope crosswires by making small adjustments of hour angle and declination;
(4) followed Venus's passage across the sky by turning the sextant slowly about its polar axis while 1st observer made his observations.

The 1st observer:
(5) directed the labourer to set lesser semicircle so that sextant was inclined in the same plane as the line joining the two objects;
(6) set movable index to approximate angular distance between Venus and Pollux;
(7) looked through movable telescope and brought image of Pollux on to telescope cross-wire by making small adjustments to lesser semicircle and index.

Entering up the observations
1. The assistant noted time by clock to nearest minute. After correction for clock error,[6] true apparent time ($18^h. 00^m.$) was set down in Column (A), in Fig. 69.
2. 1st observer took readings of movable index against the two scales on the limb. Reading of diagonal scale (49° 26' 05") set down in Column (B), the revolves and hundredth of a revolve of endless screw (1183·72) in Column (C).
3. Using conversion table, reading in Column (C) (1183·72) converted into degrees, minutes and seconds of arc (49° 25' 55") and set down in Column (D).

Historical Summary
Before June 1675: frame constructed, first of wood, later of iron, at Tower of London; index made by Thomas Tompion, the

famous clockmaker. Semicircles and other gearing made by Tower smiths; whole designed by Flamsteed (just appointed Astronomer Royal) and paid for by Sir Jonas Moore. 1676: table for converting the numbers of revolves of the perpetual screw into degrees and parts of a circle made by Flamsteed himself with sextant fixed horizontal on terrace in front of Queen's House at Greenwich; instrument then mounted in Sextant House; at that time, there was no diagonal scale, only the scale of revolves of the screw. 19 September 1676: first recorded observation. 1677: as Hooke's 10-foot quadrant proved useless, sextant used fixed in the meridian to find latitude of the observatory. 10 January 1677: storm blew roof off Sextant House; no damage to sextant.[7] December 1677: Flamsteed added diagonal scale to limb. 1680: Flamsteed started to construct 'slight' mural arc. June 1683: Edward Sylvester, Master Smith at Tower of London paid final instalment of the £120 promised by Sir Jonas Moore; references are ambiguous and could refer either to sextant or to Hooke's 10-foot quadrant.[8] 11 September 1689: first observation with final mural arc in its final form. 15 September 1690: last regular observation with sextant; by this time sextant very worn and inaccurate, but 20,000 observations had been made since 1676. 25 February 1668: last recorded observation with sextant.

1714: Board of Visitors report sextant grown very rusty. 1720: all instruments removed from Greenwich by Flamsteed's widow; the sextant has never been heard of since.

1967: full-size working model made in Museum's workshop placed on display in reconstructed Sextant House.

Contemporary Accounts

1725: John Flamsteed's *Historia Coelestis Britannica* (1725), Preface, 103–7. 1676–1705: Baily quoting Flamsteed, particularly 39–41, 127.

7.2. *Graham's 2½-foot Equatorial Sector (c. 1735)*

By George Graham of London, the first of its kind.
Probably used by Bradley at Oxford or Wanstead, 1735–42.
In Great Room, 1743–1811.
Last heard of, 1933.

Description (Figs. 71–73)

Designed by Graham specifically for measuring the positions of comets by comparison with nearby fixed stars, this was essentially a small equatorial on a 'German' mounting, on the declination axis of which was mounted towards the middle of its 2½-foot radius, a 6° sector with a 30 in. focus telescope rotating about the centre of the sector.

The instrument thus had three motions: the sector (with telescope attached) could be moved in RA about the polar axis and in Dec about the declination axis, 5-inch setting circles being provided on both axes; in addition, the telescope could be moved up to 6° on the sector arc so that differential readings of declination could be made to an accuracy of 1' of arc.

Details can be seen in Fig. 71.

Maskelyne explains how it was used at Greenwich:

> This instrument, having no peculiar room set apart for it, is used in the great room, and is removed from one window to another, according to the position of the comet may require it, and is adjusted to the meridian by means of lines drawn on the floor. Its polar axis is elevated to the altitude of the pole by the help of a peculiar apparatus contrived for the purpose. The wooden stand, on which it is supported, is also adjusted by means of a spirit-level placed in a bed fitted for receiving it in the plane of the prime vertical, lest the inequalities of the wooden floor should make the polar axis decline from the meridian in azimuth either to the right or left.[9]

Method of use (See figs. 71 and 72)

1. Decide upon reference star, which must be within 6° of declination of the comet, and as close as possible in the RA.
2. Set telescope somewhat westward of present position of star and comet. Hour circle *HI* and declination circle *LK* are fitted for setting purposes.
3. When first body comes into field of view (which could be either the star or the comet) bring onto horizontal crosswire by rotating screw Q.
4. When first body transits field, note sidereal time of transit across the three vertical wires, and reading of sector scale *N*.
5. Set telescope to declination of second body by moving screw Q. Declination clamp *O* must not be moved.
6. At second body transit note sidereal times on the 3 wires and sector scale reading as in 4 above.
7. Difference of readings between 4 and 6 above (which may have to be cleared for differences of parallax and refraction) give required RA and Dec difference.

Fig. 47. THE ALTAZIMUTH IN THE 1860s
The dome ran on cannon balls. The clock was "Graham 1". From
E. Dunkin 'A Day at the Royal Observatory', *Leisure Hour* (January
1862), 25.

Fig. 48. THE ALTAZIMUTH DOME, FROM THE SOUTH IN 1857

Fig. 49. AIRY'S ALTAZIMUTH TODAY

Fig. 50. CHRISTIE'S ALTAZIMUTH IN USE, ABOUT 1928
Frank Dyson, Astronomer Royal, is standing left. From an unidentified periodical.

Fig. 51. THE NEW ALTAZIMUTH PAVILION, ABOUT 1900
The altazimuth telescope is pointing vertically upwards with the
south collimator raised in front. The weather-vane represents Halley's
comet as seen on the Bayeux tapestry.

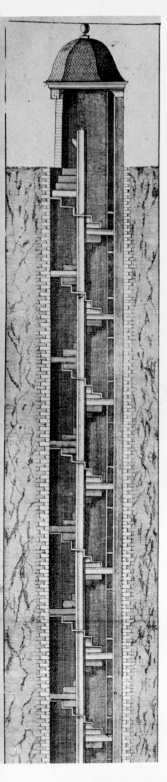

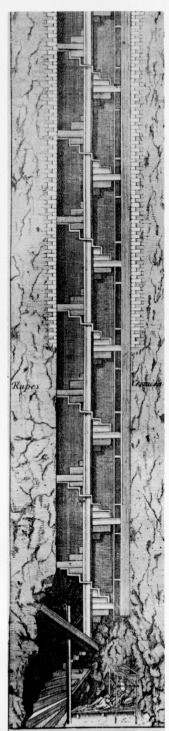

Fig. 52. "WELL OF 100-FEET EQUIPPED FOR OBSERVING THE PARAL-
LAXES OF THE EARTH", ABOUT 1676

The telescope itself seems to have been in an open wooden tube
on the right-hand side of the shaft. The object glass in its cell was
at the top of the tube. Etching by Francis Place.

*Puteus 100 pedum ad Parallaxes
Terræ observandas. p' paratus.*

Fig. 53. THE BOTTOM OF THE WELL
showing the observer on his couch, looking upwards through the
eyepiece/micrometer suspended on the plumb-line.

⊙ *The end of the axis with dot for the plumb line.*

a. The Y on which it rests.

c. The screw for regulating the plumb line.

b.b. Adjusting screws.

d.d. Iron supporters.

i.k. Brass supporter.

e. Screw & support for fastening the wood guard
  to the Brass supporter i.k.

f. The Micrometer screw.

g. A screw for taking off the pressure of the telescope
  from the micrometer screw when the instrument is not in use.

N.B. The telescope is put out of the vertical position
  in order to shew the wood guard for the plumb line.
  And the arc to the right of the telescope is broken off
  in order to shew the micrometer screw.

h. A back support on which the micrometer slides and
  to which it is clamped.

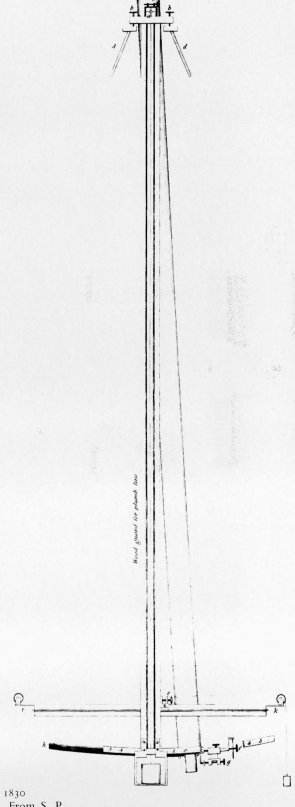

Fig. 54. BRADLEY'S 12½-FOOT ZENITH SECTOR, ABOUT 1830
Drawn by T. Taylor, Assistant at the RO, 1805–35. From S. P.
Rigaud, Bradley's Works (1832).

Fig. 55. THE EYE-END TODAY
The plumb-line can be seen bisecting a gold dot on the limb. The micrometer screw is to the right, with a knurled head. The eyepiece has a cover on.

In JANUARY M DCC LI.

DAY of the MONTH.	POINT of the LIMB.	INDEX before the Paſſage.	INDEX at the Paſſage.	INDEX after the Paſſage.	Differences in Parts of the Micrometer.	Differences reduced.	Names of the Stars.
	Deg.Min.	Rev. Sec.	Rev. Sec.	Rev. Sec.	Rev. Sec.	Min. Sec.	
♄ 19.	38 20	32 25,4	32 6,4	32 25,7	0 19,2	—0 19,0	35 Camelopardali.

After theſe Obſervations the Sector was removed into the Tranſit Room.

DAY of the MONTH.	POINT of the LIMB.	INDEX before the Paſſage.	INDEX at the Paſſage.	INDEX after the Paſſage.	Differences in Parts of the Micrometer.	Differences reduced.	Names of the Stars.
☉ 20.	32 40	29 28,5	26 2,0	29 28,8	3 26,7	—2 7,3	Capella.
♂ 22.	35 5	29 23,0	29 3,2	29 23,2	0 20,0	—0 19,8	9
	39 25	28 13,5	30 4,3	28 13,3	1 25,0	+0 58,4	γ
	35 50	28 15,5	32 24,4	28 15,3	4 9,0	+2 23,4	α
	34 15	28 30,8	30 2,2	28 31,2	1 5,2	+0 38,8	↓
	33 50	29 5,7	34 29,3	29 6,0	5 23,4	+3 11,7	δ ⟩ Perſei.
	36 50	29 4,0	32 28,0	29 3,5	3 24,3	+2 4,9	A
	36 35	29 24,5	28 5,0	29 23,7	1 19,0	—0 52,4	λ
	34 40	29 2,3	29 9,0	29 2,8	0 6,5	+0 6,4	μ ½ paſt.
	32 50	28 15,5	24 10,7	28 15,5	4 4,8	—2 19 3	d
	40 15	26 20,7	27 : 24,8	26 21,0	1 3,9	+0 37,5	1
	39 50	27 6,4	31 12,2	27 6,8	4 5,6	+2 20,1	2 ⟩ Camelopardali.
	43 10	27 27,2	31 27,4	27 26,9	4 0,4	+2 14,9	4
	40 15	27 21,3	26 25,5	27 20,7	0 20,5	—0 20.2	7
A	B	C	D	E	F		

Fig. 56. PUBLISHED ZENITH–SECTOR OBSERVATIONS FOR 1751
From Bradley, *Astron. Obs.*, I, 102.

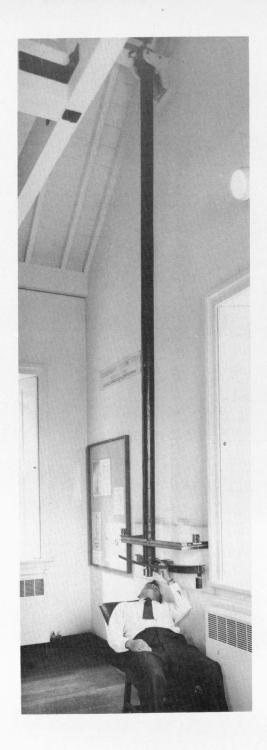

Fig. 57. THE ZENITH SECTOR TODAY

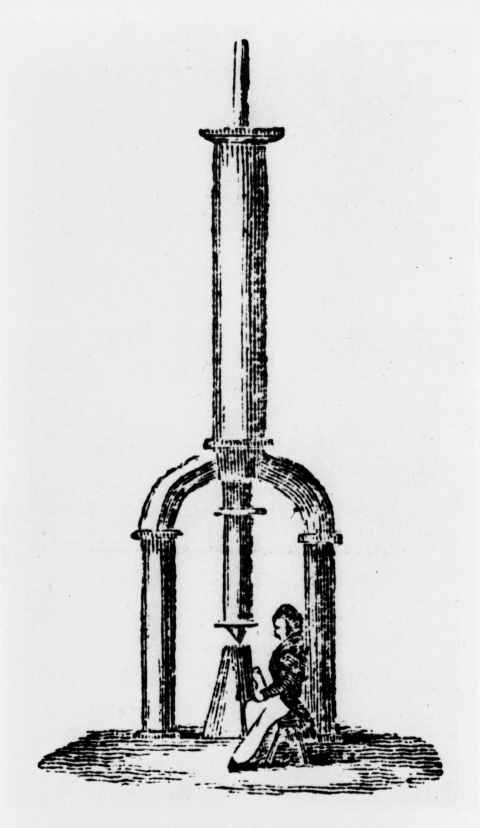

Fig. 58. THE GREAT ZENITH SECTOR IN 1835
From *The Weekly Visitor* (17 February 1835), 584.

OBSERVATIONS with the TWENTY-FIVE FEET ZENITH TELESCOPE.

Date.	Star.	Approx. A.R.	Thermometer.		Face of Inst.	Reading of the Instrument.			Ob-servers.	Remarks.
			Upper.	Lower.		Rev^tns of Screw.	Parts.	Vern.		
1833.		H. M.								
Sept. 12	✳ Cygni - - - -	20 31	55	55	W	60	85	28	F. S.	
	✳ - - - -	21 39	54	,,	,,	93	32	50	,,	Very tremulous.
	3 Lacertæ - - -	22 17	53	51	,,	77	49	00	,,	
13	γ Draconis - - -	17 52	57	55	E	83	29	11	H.	
	✳ - - - -	20 31	,,	,,	,,	117	49	60	F. S.	
	✳ - - - -	21 39	,,	54	,,	85	07	10	,,	
	3 Lacertæ - - -	22 17	53	,,	,,	100	86	70	,,	
14	✳ - - - -	21 39	59	57	W	93	29	20	F. S.	
	3 Lacertæ - - -	22 17	,,	,,	,,	77	51	18	,,	
15	ι Cygni - - - -	19 25	59	56	W	75	18	73	W. R.	
	✳ - - - -	20 0	,,	,,	,,	73	00	00	,,	
	✳ - - - -	20 30	,,	,,	,,	60	85	00	E.	
	✳ - - - -	21 39	,,	55	,,	93	29	30	F. S.	
	3 Lacertæ - - -	22 17	55	,,	,,	77	50	70	,,	
16	γ Draconis - - -	17 52	57	57	W	95	03	75	H.	
	3 Lacertæ - - -	22 17	56	56	,,	77	56	70	F. s.	Cloudy.
17	ι Cygni - - - -	19 25	57	57	E	103	11	34	H.	
	3 Lacertæ - - -	22 17	56	56	,,	100	77	45	F. s.	

Fig. 59. PUBLISHED ZENITH TELESCOPE RESULTS FOR 1833
By Pond before the considerable modifications in 1837. From *Astron. Obs.*, 1833.

Fig. 60. AIRY'S REFLEX ZENITH TUBE—OPTICAL PRIN-
CIPLES
From J. N. Lockyer, *Stargazing* (1878).

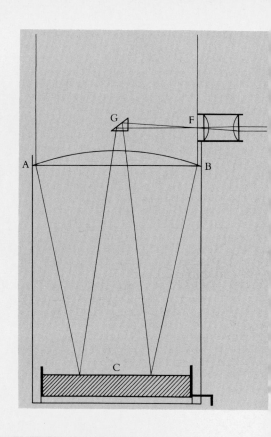

OBSERVATIONS of γ DRACONIS with the Reflex Zenith Tube, and Reduction of the Observations—*continued.*

Day and Hour of Observation, 1863.	Observer.	Position of Mic. A.	Wire used.	Micrometer Readings. A	Micrometer Readings. B	Level Readings. div.	Level Readings. div.	Equivalent for Level.	Sum of Equivalents for Wire, for Micrometer-Readings, and for Level Readings.	Assumed Instrumental Constant.	Star's Z.D. North from Observation.	Correction to Mean Z.D. North for 1863, Jan. 1.	Mean Zenith Distance North, 1863, Jan. 1, from each Observation.	Concluded Mean Zenith Distance North, 1863, Jan. 1.
				r	r	div.	div.	"	"	"	"	"	"	"
d h														
August 14. 8	N	Left	16	52·464	34·505	15·0	69·2	0·55	335·49	214·84	120·65	−19·19	101·46	101·84
" "		Right	15	51·020	34·505	28·4	83·0	0·72	93·43		121·41		102·22	
August 15. 8	JC	Right	15	50·997	34·506	28·6	83·3	0·72	93·06		121·78	−19·38	102·40	104·15
" "		Left	16	50·997	36·248	14·3	69·0	0·54	340·11		125·27		105·89	
August 17. 8	E	Left	16	51·000	36·128	14·0	70·0	0·54	338·15		123·31	−19·76	103·55	102·14
" "		Right	15	49·392	36·128	27·6	83·5	0·72	93·35		121·49		101·73	
August 19. 8	JC	Right	15	49·114	36·324	27·2	84·4	0·72	91·96		122·88	−20·14	102·74	104·78
" "		Left	16	49·114	38·232	12·8	70·7	0·54	341·80		126·96		106·82	
August 20. 8	C	Left	16	49·116	38·155	12·8	70·7	0·54	340·56		125·72	−20·28	105·44	104·67
" "		Right	15	47·205	38·155	26·6	85·8	0·73	90·67		124·17		103·89	
August 24. 8	C	Left	16	47·050	40·239	13·7	70·5	0·54	340·85		126·01	−20·84	105·17	104·87
" "		Right	15	45·048	40·239	27·5	84·5	0·72	89·44		125·40		104·56	
September 3. 7	K	Left	16	45·111	42·389	12·0	68·4	0·52	344·37		129·53	−22·04	107·49	107·49
September 17. 6	E	Left	16	45·088	42·388	13·5	70·7	0·54	343·99		129·15	−22·99	106·16	105·25
" "		Right	15	42·785	42·388	26·2	83·4	0·71	87·52		127·32		104·33	
September 18. 6	C	Left	16	42·784	44·606	14·0	71·4	0·55	342·55		127·71	−23·04	104·67	104·16
" "		Right	15	40·594	44·606	40·5	97·6	0·89	88·15		126·69		103·65	

Fig. 61. PUBLISHED ZENITH–TUBE RESULTS FOR 1863
From Airy, *Astron. Obs.*, 1863, 57.

'B' micrometer

7

6

5

4

3

2

1

8

9

'A' micrometer

10

11

1.	Support tube	6.	O.G.
2.	Clamp for focus	7.	Micrometer carriage 'B' end
3.	Focusing knob	8.	Prism reflector
4.	Rotating head carrying O.G. and micrometers	9.	Spirit level
5.	Fixed eye piece	10.	Inner focusing tube carrying 4
		11.	Counter balance supporting 10

Fig. 62. AIRY'S REFLEX ZENITH TUBE

1	Mercury trough	6	Clamp
2	Float	7	Sector
3	Telescope	8	Setting circle
4	Trunnions	9	Plate holder
5	Indexing device		

Fig. 63. COOKSON'S FLOATING ZENITH TELESCOPE AT CAMBRIDGE
IN 1900
From *MNRAS* LXI 5 (1901), Plate 7.

Fig. 65. PHOTOGRAPHIC ZENITH TUBE–GENERAL ARRANGEMENT

Fig. 65. PHOTOGRAPHIC ZENITH TUBE–GENERAL ARRANGEMENT
The OG L is mounted on the rotary R; M is the mercury bath, P the photographic plate, and T the conical tube supporting the rotary, which turns in the ball race B. From *Occasion Notes of the RAS*, 21 November 1959, 228—by permission.

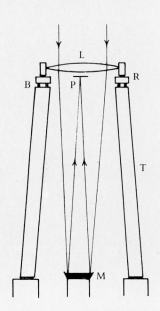

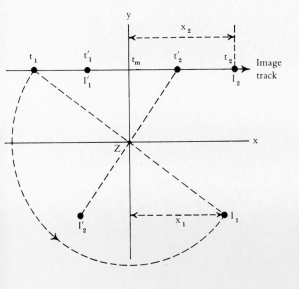

Fig. 66. THE PZT PHOTOGRAPHIC RECORD
The x axis is the projection of the prime vertical, the y axis that of the meridian. The star image moves along the track, its position at the four times being as shown. t_m is the time of meridian transit. The rotary photographic plate is reversed between each exposure and the subsequent one. After the fourth exposure, the four images on the plate are situated at I_1, I_1', I_2' and I_2.

The perpendicular distance between the lines $I_1 \, I_2'$ and $I_1' \, I_2$ is a measure of twice the meridian zenith distance of the star. The scale value of the plate is obtained from the distances $I_1 \, I_2'$ or $I_1' \, I_2$ and the observed times.

The slight curvature of the image track is not shown in the diagram. From *Occ. Notes R. Astron. Soc.*, *RAS*, 21 November 1959), 229—by permission.

Fig. 67. THE PZT CONTROL DESK AT HERSTMONCEUX

Fawes Sextantis Anterior 7 ped: Rud:.

Fawis Sextantis Posterior 7 ped: Rud:.

Fig. 68. FLAMSTEED'S 7-FOOT SEXTANT, FRONT AND BACK, ABOUT 1676

This is the only one of the set of Francis Place engravings whose descriptive text is known—in *Hist. Cel. Brit.*, III (1725), 103–7. Note the run-off roof in the right-hand picture which looks east, which has been omitted from the other picture. The Sextant clock can be seen in the shadowed part of the left-hand picture which looks west.

ANNO CHRISTI MDCLXXXIV.			VENERIS DISTANTIÆ à Sole & Fixis.		Diſtantiæ per lineas diagonales.			Diſtantiæ per Cochleæ Revolutiones & Circuli partes Reſpondentes.			
Menſe, Die. Styl. Vet.		Temp. App. H. M.			°	′	″	Revol. Cent.	°	′	″
☽ Sept. 15 vel ☿ 16 mane.	17	27	Venus ——— à Polluce ———		49	26	15	1183 75	49	26	00
		29		rep.	49	26	25	1183 82	49	26	10
		55		iter.	49	26	40	1183 94	49	26	28
		58		denuò	49	26	45	1183 98	49	26	35
♀ 19	22	46	Venus ——— à limbo Solis proximo ———		31	28	50	753 51	31	28	36
		48		rep.	31	28	45	753 53	31	28	30
		50	à limbo Solis remoto ———		32	00	45	766 25	32	00	28
		52		rep.	32	00	50	766 29	32	00	35
☉ 21	16	56	Venus ——— à Polluce ———		49	24	15	1182 87	49	23	48
	18	00		rep.	49	26	05	1183 72	49	25	55
		01		iter.	49	26	05	1183 72	49	25	55
	23	38	à limbo Solis proximo ———		33	08	05	793 15	33	07	45
		40		rep.	33	08	15	793 22	33	07	55
		42	à limbo Solis remoto ———		33	40	00	805 93	33	39	42
		43		rep.	33	40	00	805 93	33	39	42
			Vide eodem Die Solis & Veneris Merid. à Vertice Diſtantias, loco proprio.								
☿ 25	22	35	Venus ——— à limbo Solis remoto ———		36	28	55	873 42	36	28	30
		37		rep.	36	29	00	873 47	36	28	37
		40	à limbo Solis proximo ———		35	57	10	860 68	35	56	38
		41		rep.	35	57	20	860 71	35	56	42
			Vide Solis & Veneris à Vertice eodem Die Obſervationem, proprio loco.								

A   B   C   D

Fig. 69. PUBLISHED SEXTANT RESULTS FOR 1684
From *Hist. Cel. Brit.*, I, 180.

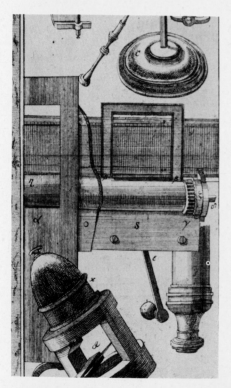

Fig. 70. DETAILS OF FLAMSTEED'S SEXTANT

Left—Front of index of moving telescope, showing diagonal scale on limb, two fiducial edges, telescope eyepiece and operating handle ε.
Right—Back of index, showing operation of perpetual screw, worked by handle d. The wormwheel could be withdrawn clear of the threads on the limb by screw F to permit initial setting of telescope. Polar axis *a* (shown vertical), showing method of operation of greater (declination) semi-circle worked by handle A, and lesser (inclination) semicircle worked by handle k Flamsteed's own description has not survived. Detail from Francis Place etching *Partes Instrumentorum* . . . , about 1676.

Fig. 71. GRAHAM'S 2½-FOOT EQUATORIAL SECTOR
From Rees' *Cyclopaedia*, *Astron. Insts.*, Plate XIII dated 1811.

OBSERVATIONS of the Difference of Right Ascension and Declination between the COMET of 1769 and several fixed Stars, made with the 30 Inch Equatorial Sector of Graham's Construction.

DAY of the MONTH.	Objects observed.	Transit of Star and Comet.			N° of Degrees, &c. by Arch of Sector.	Red. of Clock to Transit Clock.	Sidereal and mean Time at Transit of Comet.
		1st Wire.	2d Wire.	3d Wire.			
1769		M S	M S	M S	D M S		H M S
SEP. 3	Comet · · · · · · ·	20 52	0 21 55	22 57	3 20 0	+ 1 46	S. T. 0.24.20,5
	γ Orion · · · · · ·	24 33½	0 25 36	26 38	3 24 50	· · · ·	M. T. 13.30.36
	Comet prec. Star · · ·	3 41½	3 41	3 41	0 4 50N		
W. B. {	Comet · · · · · · ·	34 38	0 35 42	36 45	3 20 0		S. T. 0.38.7,5
	γ Orion · · · · · ·	38 11+	0 39 13½	40 16	3 25 5	· · · ·	M. T. 13.44.20
	Comet prec. Star · · ·	3 33+	3 31½	3 31	0 5 5N		
	Comet · · · · · · ·	47 5	0 48 8	49 11	3 20 0	· · · ·	S. T. 0.50.33,5
	γ Orion · · · · · ·	50 29	0 51 31½	52 34	3 25 0	· · · ·	M. T. 13.56.44
	Comet prec. Star · · ·	3 24	3 23½	3 23	0 5 0N		
SEP. 4 W. B. {	A Orion · · · · · ·	29 34	1 30 36½	31 38+	3 20 0	+ 1 46	
	Comet · · · · · · ·	37 4	1 38 6½	39 10	3 46 25	· · · ·	S. T. 1.40.33.7
	Comet follows Star · ·	7 30	7 30	7 32	0 26 25S	· · · ·	M. T. 14.42.40

Fig. 72. GRAHAM SECTOR RESULTS FOR 1769
From Maskelyne, *Astron. Obs.*, 1769, 67.

Fig. 73. ANCIENT INSTRUMENTS IN THE TRANSIT CIRCLE ROOM ABOUT 1898

Graham's equatorial sector, which has since disappeared, is at the bottom, slightly left of centre. Hanging on the wall is Halley's quadrant (clamp and micrometer since disappeared) and Bradley's zenith sector. The high survival rate of Greenwich instruments owes much to the practice of hanging superseded instruments on the walls of the Transit or Transit Circle Room, a practice started by Pond and continued by his successors.

ig. 74. MORE ANCIENT INSTRUMENTS, ABOUT 1898

he north-west corner, showing three transit instruments, four strid-
ng levels, Bradley's setting-semicircle and a barometer by Nairne.

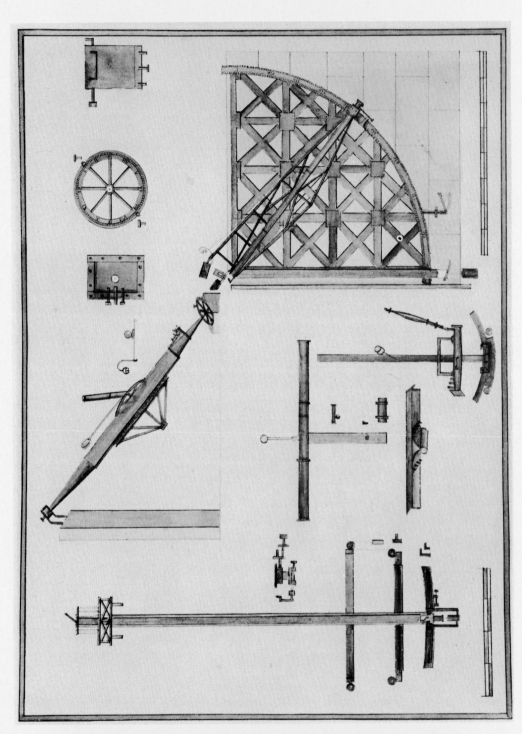

Fig. 75. SISSON'S EAST SECTOR (*centre*) BRADLEY'S ZENITH SECTOR (*left*),
BRADLEY'S BRASS QUADRANT (*right*), SKETCHED ABOUT 1785

Another of Charnock's drawings (see Fig. 31). It is very tantalizing that no   sector, then mounted in the East Dome at the end of the north terrace of

OBSERVATIONS of the COMET of 1815

WITH THE FIVE-FEET EQUATORIAL SECTOR

IN THE WESTERN DOME.

1815	Sidereal Time.	Mean Time.	Comet precedes or follows the Star in Time.	Ditto in Parts of the Equator.	Comet N. or S. of Star in Declination.	Name and Character, and Right Ascension and Declination of the Star nearly.	Remarks.
	h ′ ″	h ′ ″	′ ″	° ′ ″	′ ″		
MAY 22	13 54 6	9 55 25	prec. 16 56	4 14 0	S. 10 24	*Telescopic* ✳ ℞ 133° 8′ 15″ Decl. 61 12 7 N.	
23	13 44 55	9 42 19	prec. 9 2	2 15 30	S. 20 35	Same Star.	
24	13 44 25	9 37 52	prec. 1 11	0 17 45	S. 32 37	Same Star.	
25	13 42 54	9 32 25	foll. 6 37	1 39 15	S. 46 31	Same Star.	
27	14 5 8	9 46 43	prec. 23 8	5 47 0	N. 2 0	*v Urf. Maj.* (Bode.) ℞ 144° 26′ 30″ Decl. 59 54 2 N.	

g. 76. PUBLISHED RESULTS FROM SISSON SECTOR, 1815
om Pond, *Astron. Obs.*, 1815, 242.

Fig. 77. THE ORIGINAL SECTOR PART OF SISSON'S
5-FOOT EQUATORIAL SECTOR,
removed when the instrument was modified in
1780. This is the only part of Sisson's two sectors
to have survived.

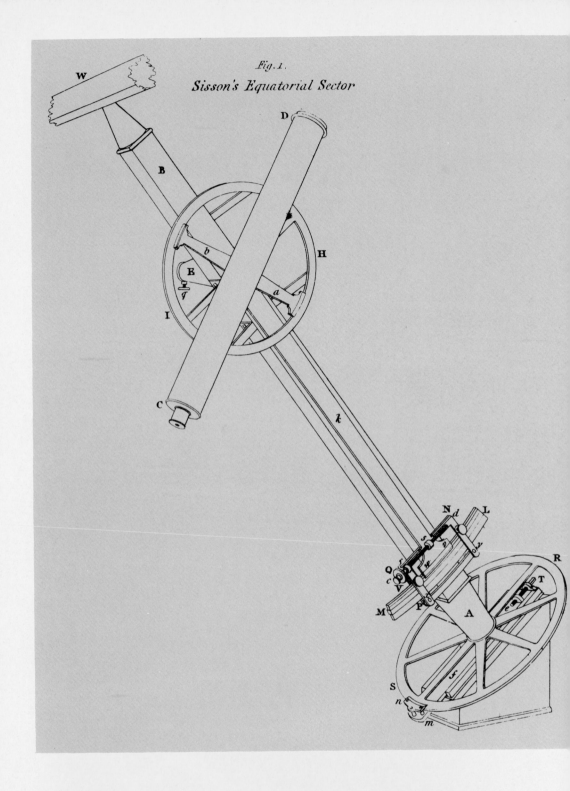

Fig. 78. SISSON'S SECTOR AFTER MODIFICATION IN 1780
Though this engraving purports to show the Greenwich sector, it
conflicts in some details—the upper end of the polar axis, for example
—with Charnock's drawing in Fig. 75, which is likely to be the
more authentic. From Rees, *Cyclopaedia*, Plate dated 1811.

Fig. 79. THE SHUCKBURGH 4.1-INCH REFRACTOR
In 1967, at the Science Museum, London. In the background is
the South/Smyth/Lee 6-inch refractor.

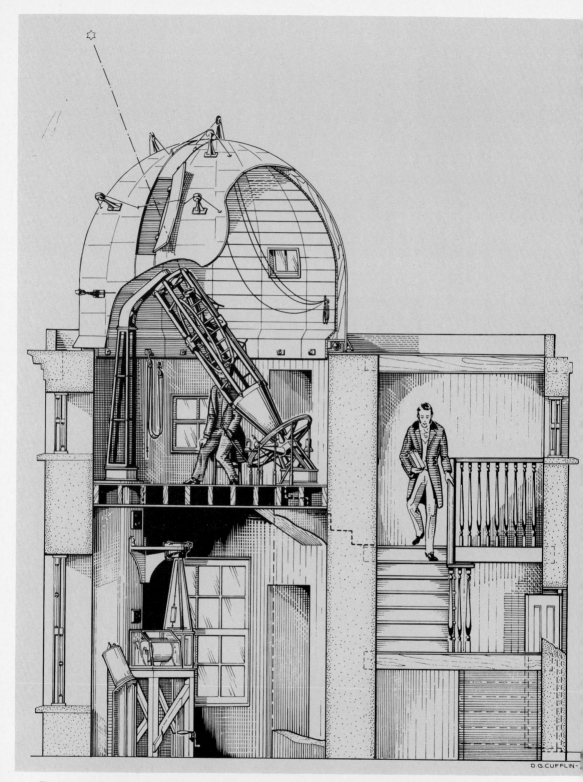

D.G.CUFFLIN

Fig. 80. THE NORTH–EAST DOME ABOUT 1860,
looking east, showing the dome erected for Sisson's
equatorial in 1773, the Shuckburgh equatorial which
replaced the Sisson in 1816, and Airy's Chronograph
installed on the ground floor in 1852. The Shuckburgh
piers are incorrectly shown. They were actually of stone,
as in Figs. 75 and 79. Drawn by the late Donald Cufflin
in 1971.

Fig. 81. THE SHEEPSHANKS 6.7-INCH REFRACTOR IN THE ALTAZI-
MUTH PAVILION IN 1966

Fig. 82. THE 12.8-INCH GREAT EQUATORIAL ABOUT 1860
From E. Dunkin, *The Midnight Sky* [new edition 1880], 236.

Fig. 83. THE SHEEPSHANKS *(centre)* AND GREAT EQUATORIAL *(right)*
DOMES IN 1880
The Reflex Zenith Tube Room can be seen, under the Sheepshanks
Dome in the re-entrant angle between the Transit Circle Room
and the old Transit Room (whose roof openings can still be seen
left, now closed to the weather). From a water-colour signed *July
1880 Christabel Airy* in the possession of the RGO.

Fig. 84. PHOTOHELIOGRAPH ERECTED IN TRANSIT-OF-VENUS PORTABLE OBSERVATORY 1874
From N. Lockyer *Stargazing past and present* (1878), p. 461.

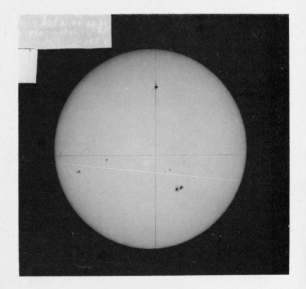

Fig. 85. PHOTOGRAPH OF SUN FROM 6-INCH GLASS PLATE TAKEN IN 1883
From 1884, 10-inch plates were used.

Fig. 86. THE ASTROGRAPHIC TELESCOPE AT GREENWICH IN 1904
From W. Christie, *Astrographic Catalogue . . .* (1904).

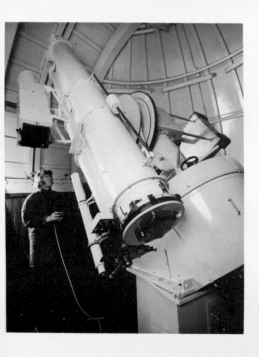

Fig. 87. THE ASTROGRAPHIC TELESCOPE AT RGO
HERSTMONCEUX IN 1974,
on the new mounting of 1969. Photo by David Calvert.

Fig. 88. THE NEW 36-FOOT SOUTH-EAST EQUATORIAL DOME UNDER
CONSTRUCTION IN 1893
This onion-shaped dome replaced the drum-shaped dome seen in
Fig. 83 when the 12.8-inch refractor was replaced by the 28-inch
refractor—which was 10 feet longer than its predecessor.

Fig. 89. THE SOUTH-EAST EQUATORIAL DOME ABOUT 1935,
with the 28-inch Refractor; from the Middle Garden. The cladding
was of papier-mâché.

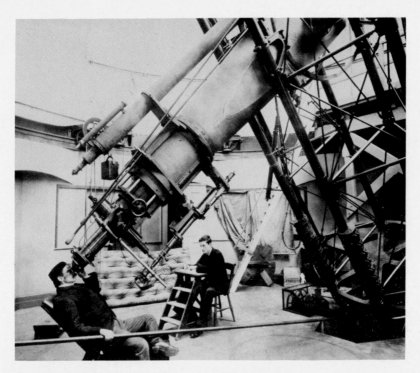

Fig. 90. THE 28-INCH VISUAL REFRACTOR, ABOUT 1899
Mr. T. Lewis at the eyepiece, Mr. W. Bowyer recording. The half-prism spectroscope mounted under the main tube was later dismounted.

Fig. 91. THE 28-INCH TELESCOPE AND POLAR AXIS
RETURN TO GREENWICH IN 1971
The telescope and mounting had been sent to Herstmonceux in 1947.

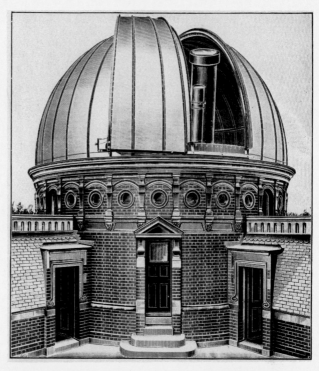

Fig. 92. THE 26-INCH REFRACTOR IN THE 30-FOOT THOMPSON
DOME, ABOUT 1900
From a catalogue of T. Cooke & Sons of York, makers of the
dome.

Fig. 93. THE 26-INCH PHOTOGRAPHIC REFRACTOR AT
GREENWICH ABOUT 1925
The Merz 12.8-inch refractor in a steel tube is mounted
as guiding telescope above the main tube, the Thompson
9-inch photoheliograph is mounted below. The 30-inch
reflector was at that time mounted as counterpoise, off
the picture to the left (Fig. 95).

Fig. 94. THE 26-INCH REFRACTOR AT HERSTMONCEUX IN 1974

The Merz refractor is still mounted as guiding telescope but the thirty-inch reflector has been removed and a weight substituted at the other end of the declination axis. Photo by David Calvert.

Fig. 95. THE 30-INCH PHOTOGRAPHIC REFLECTOR ABOUT 1930

Fig. 96. THE 30-INCH REFLECTOR AT HERSTMONCEUX IN 1974,
on the fork mounting of 1956. Photo by David Calvert.

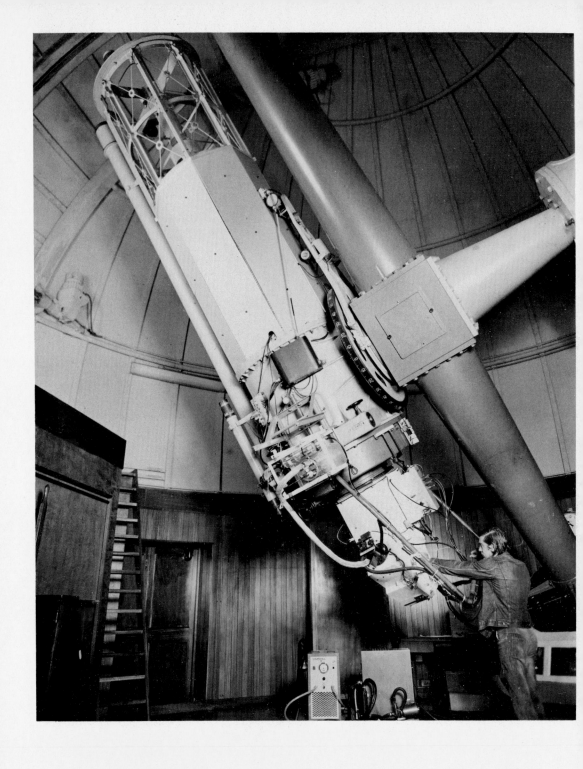

FIG. 97. THE YAPP 36-INCH REFLECTOR AT HERSTMONCEUX IN
1974,
with RGO image-intensifier spectrograph and photometer. Photo
by David Calvert.

Fig. 98. THE YAPP DOME AT GREENWICH ABOUT 1934,
in the Christie enclosure in the park (originally the magnetic observatory).

Fig. 99. THE ISAAC NEWTON 60-FOOT DOME, HERSTMONCEUX
IN 1974
Photo by David Calvert.

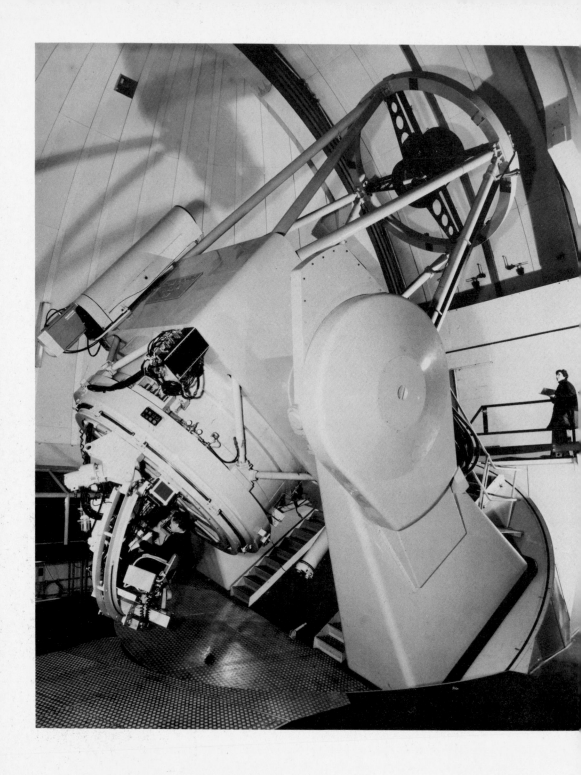

Fig. 100. THE 98-INCH ISAAC NEWTON REFLECTOR IN 1974
The observer can be seen, left, in a chair at the Cassegrain focus.
He can also observe from the prime focus. There is a separate control
console. Photo by David Calvert.

Historical Summary

1738: Graham's sector described in Smith's *Opticks*; we have no details of how Bradley acquired this instrument nor any record of observations before 1742. 1742: Bradley comes to Greenwich, apparently bringing sector with him; 1748–9: comet observations in Great Room. 1750: sector became part of observatory apparatus, Board of Ordnance paying £35 for it. 1756 and 1769: comet observations in Great Room. 1770: Maskelyne complains of Graham sector; because telescope was of 'the old sort', faint comets were difficult to see; because it had to be used in Great Room, it had to be moved from window to window, high comets could not be seen and it lacked stability; he asked for a larger sector in purpose-built observatory; in the event, the summer houses were converted and two 5-foot sectors ordered from Jeremiah Sisson. 1771 and 1773: comet observations in Great Room. 1779: last observations with Graham sector; first observation with Sisson sector.

1864: examined by Airy; placed as relic in Transit Circle Room.

1933: in the South Building, Lower Museum, Case G; its present whereabouts is not known.

Contemporary Accounts

1738: Smith, *Opticks*, 350–54, Pl. 51, Figs. 608–12. 1776: Maskelyne, *Astron. Obs.*, I, xii. 1790: Vince, 139–41, Pl. V, Fig. 47.

7.3. *Sisson's 5-foot Equatorial Sectors* (1773)

By Jeremiah Sisson of London.

Surviving part of one sector signed *Jeremiah Sisson 1773*.

Description (Figs. 75–78)

These two identical instruments were mounted in the new observatories—converted from Wren's summer houses at each end of the north terrace of Flamsteed House in 1773. No illustration or detailed description of them in their original form survives but we can infer from the surviving sector-part (Fig. 77) that they were basically scaled-up versions of Graham's sector just described, but mounted on a long polar axis in the 'English' fashion—as in the centre section of Fig. 75.

The 5-foot 20° sectors were mounted on the polar axis 3 feet in from their centres. The declination circle was 2 feet in diameter, the hour circle probably of the same radius. The 5-foot focus achromatic telescope was of 4-inch aperture.[10]

These two sectors were severely criticized for, among other things, the weakness of their polar axes and the smallness of their circles. In 1780 the eastern sector was modified to Maskelyne's design as in Fig. 78. But money was short: apparently using the old polar axis and telescope, the old sector-part was discarded in favour of a long index free to move 10° either side of the polar axis. Whether the circles were enlarged we do not know but it seems doubtful. Sisson was paid £89 9s. for the alterations.[11]

About 1789 the western sector was "altered and made a serviceable instrument with the least expense possible" by Troughton for £25.[12] Once again details are lacking but we do know that the 10-foot 3-inch polar axis, 2-foot circles and 5-foot telescope were retained.

About 1827, George Dollond replaced the 5-foot telescope by one of 30 inches focus with an OG by Tulley, mounting it as an ordinary equatorial but retaining the old polar axis and circles.[13]

Historical Summary

1770: Maskelyne asks for replacement for Graham sector; decision to modify the two summer houses instead of building new observatory meant that two sectors were needed because field of view was restricted by Great Room. 1772: order placed with Jeremiah Sisson for two 5-foot sectors at £300 and with Dollond for two 4-inch OG's at £21; work started on converting summer houses. 1774: sectors delivered. September 1775: Maskelyne reports imperfections "through the neglect of the instrument maker". 1776: west sector removed by Ramsden for examination and modification. 1778: new equatorial instrument with 5-foot telescope and 5-foot circles ordered from Ramsden; because Ramsden was so slow and because of disputes over money this order was cancelled about 1784; almost certainly some parts of it were incorporated in the equatorial with 4-foot circles bought by Sir George Shuckburgh and eventually installed in the East Dome in 1818 (see p. 87). 1780: east sector modified by Sisson to Maskelyne's design as described above; old sector parts removed; new long index. April–October 1781: observations of 'Georgian planet', Uranus, discovered by Herschel 13 March; in this period, the east sector was moved from dome to dome as needed. 1788: west sector modified by Troughton "with least expense possible" (£25); no details.

June 1816: east sector dismantled; Shuckburgh equatorial installed in East Dome. 1824: old east sector lent to Herschel and South who incorporated polar axis and circles into their 7-foot equatorial.[14] c. 1827: west sector: 5-foot telescope replaced by one of 3-inch focus by Tulley. 1846: west sector: all parts except telescope sent to Cape Observatory as mounting for old 46-inch Dollond telescope.

Contemporary Acccount
   1790: Vince, 141–8, Fig. 48.

NOTES

1. Baily, p. 128.
2. *HC* I, 390–5.
3. *HC* III, 103.
4. Baily, p. 127.
5. There is a sketch of the sextant in this mode in observation book for 8 May 1677 (RGO MS. 1/60).
6. Until comparatively recently, astronomers started counting their hours from noon. Thus 18^h 00^m on 21 September to the astronomer was 6 o'clock in the morning of 22 September of the ordinary citizen.
7. RGO MS. 36/37, Flamsteed to Moore, 11 January 1677.
8. PRO/WO 47/63(61) of 24 March 1683 and WO55/470 (154) of 9 June 1683.
9. Maskelyne, *Astron. Obs.*, I (1776), xii.
10. Maskelyne, *Astron. Obs.*, IV, Quad. 56.
11. RS MS. 371/5.
12. ROV Vol. II, 16 November 1787.
13. Airy, *Astron. Obs.*, 1845.
14. *Phil. Trans.*, 1825, p. 12.

8

The Large Equatorials

A<small>N</small> equatorial telescope is one that is mounted so that it can rotate about a polar axis parallel to the axis of the Earth. With such a telescope it is possible to follow the apparent motion of the heavenly bodies across the sky by simply rotating it about this polar axis.

The idea of such a mounting occurred long before the invention of the telescope. Ptolemy's *armilla* of the first century A.D., Regiomontanus's *torqueta* of the 1460s and Tycho's *equatoria* of the 1580s were all equatorially mounted, albeit with open sights. But the first equatorially mounted telescope is ascribed to Scheiner in 1620, while the first large instrument with telescopic sights so mounted seems to have been Flamsteed's sextant of 1676, described above.

At least two other telescopes of great significance in the history of the development of this type of mounting have close connections with Greenwich—Abraham Sharp's double telescope with 6-foot and 2-foot tubes, of about 1700, probably the world's earliest surviving telescope on a portable equatorial mounting, the property of the Yorkshire Philosophical Society, now preserved in the National Maritime Museum;[1] and Jesse Ramsden's 4·1-inch Shuckburgh telescope of 1791, the world's first large equatorial on a fixed mounting, used at Greenwich from 1818 to 1929 and now preserved in the Science Museum, London. There are three main variations in the form of this mounting—the 'English', the 'German' and the 'Fork' types.

The English mounting has a long polar axis supported top and bottom by the north and south piers, the axis either taking the form of a yoke with the telescope swinging inside it (as in the Shuckburgh equatorial), or is solid, with the telescope mounted to one side (as with Sisson's sector described on pp. 81–2 above), usually with a counterpoise on the other side.

The German, or Fraunhofer, mounting—the Sheepshanks equatorial for instance—has but one pier, on top of which lies the polar axis. The declination axis forms a 'T' with the polar axis, the telescope being mounted on one side of it at one side of the pier, a counterpoise on the other (on the Thompson equatorial the counterpoise took the form of a second telescope).

In the Fork mounting—e.g. the 98-inch Isaac Newton telescope—the polar axis is supported at the bottom only, being fork-shaped with the telescope swinging in the fork.[2]

The classical English equatorial suffers from the disadvantage that the polar regions are inaccessible; with the German equatorial the telescope has to be moved from one side of the pier to the other in order to cross the meridian; the fork mounting generally requires much heavier engineering.

The terms 'English' and 'German' equatorial seem to have been used first by G. B. Airy in his description of the Northumberland equatorial at Cambridge.[3]

Rotation about the polar axis was performed by hand with the early instruments but, since about the 1830s, mechanical driving-clocks have generally been fitted, driving the telescope at a rate which compensates for the Earth's diurnal motion.

Robert Hooke described such a device, controlled by a conical pendulum, as early as 1674[4] but the first driving-clock in England is said to be that designed by Sheepshanks fitted to the South/Smyth/Lee 6-inch equatorial of the 1820s, described on p. 118.[5]

8.1. *Shuckburgh 4·1-inch Refractor* (1791)
The World's first large Equatorial Telescope

Made by Jesse Ramsden for Sir George Shuckburgh, Bart. (1751–1804).

Signed on hour circle: *Ramsden London/Redivided by Troughton & Simms, 1860.*

Mounted at Shuckburgh, Warwickshire, 1791–c. 1810.

Presented to RO by Shuckburgh's heir, 1811.

Mounted in N.E. Dome at Greenwich 1816–1929.

Description (Fig. 79)

Telescope: 4·1-inch aperture achromatic doublet 6-inch focal length, greatest magnification 400.

Mounting: English, with declination axis in centre of yoke. Polar axis 8 feet 4 inches long.

Circles 4 feet diameter, divided by 10' reading to seconds by micrometers.

Driving Clock: none.

Main Apparatus: divided eyepiece micrometer by Ramsden. Micrometer by Dollond.

Astronomical clocks: at Shuckburgh, Arnold degree clock (see p. 136). At Greenwich to 1874, 'Arnold 1'.

Dome at Greenwich: hemispherical, 11 feet diameter.

Inscriptions: according to Shuckburgh's description[6] there was a plaque on the polar axis cone with the following Latin inscription:

> This *panorganon* ('measurer of the heavens'), conceived and now after ten long years completed by Jesse Ramsden of London, the celebrated optician, who was by far the greatest practitioner of his art, was commissioned in 1791 by Sir George Shuckburgh as a token of his love for astronomy and for the encouragement of that science.

Today, that plaque has been replaced by another, also in Latin, saying that this equatorial was placed at Greenwich by the generous wish of Shuckburgh's heir, Charles Cecil Cope Jenkinson, Esq., 1811.

Though of great historic interest as the world's first large equatorial, this telescope was, alas, never a success. Pearson said of it in 1829:

> "... the framework was found too slender for the length of the polar axis, and it remains at Greenwich, a proof of its former proprietor's munificence rather than of its maker's success in the strength and stability of its essential parts."[7]

Airy was even more scathing 25 years later. In making the case for the new Great Equatorial, he tells the Visitors that at least £100 would be needed to make the Shuckburgh serviceable:

> "And when this is done, we should have a small indifferent telescope, on a weak frame, in a position bad beyond anything that an astronomer ever imagined."[8]

Purpose

Eclipses, occultations, etc. Comet positions.

Historical Summary

1781: ordered by Sir George Shuckburgh from Ramsden. Almost certainly, Ramsden used parts of the equatorial ordered for the RO in 1778 and cancelled in 1784. (But 5-foot circles were specified for the Greenwich telescope while Shuckburgh's had 4-foot circles (see p. 82 above)). By 1793: mounted at Shuckburgh, Warwickshire, as described in *Phil. Trans.*, 1793, 67–128.

1811: presented to RO by Shuckburgh's heir, the Hon. Cecil Jenkinson (later Lord Liverpool). Delivered August 1812. 1813–5: planned to be mounted as altazimuth similar to Palermo circle on new pier east of Circle Room (later Sheepshanks dome) but mounting proved so unsteady that telescope was never so mounted.[9] June 1816: mounted as an equatorial by Matthew Berge in place of Sisson Sector, in the East Dome probably using existing stone piers. A most unfavourable situation as S.W. sky is hidden by Octagon Room. (See Airy's remarks on p. 8). 1824: only two observations recorded before 1824. 1838: declination circle re-divided because Berge's old divisions had been worn away. 1860: hour circle re-divided by James Simms.

November 1929: dismounted and transferred to Science Museum, London.

Contemporary Accounts

1793: Sir George Shuckburgh Bart, "An account of the Equatorial Instrument", *Phil. Trans.*, 1793, 67–128.

1836: Airy, *Astron. Obs.*, 1836, iii.

8.2. *West Dome 30-inch Equatorial (c. 1827)*

By Dollond, using OG by Tulley and mounting of old equatorial sector by Sisson.

In West Dome, c. 1827–46.

No parts are known to have survived.

Description

Telescope: 3·5-inch aperture, 30-inch focal length, taken from Tulley's small telescope in Great Room.

Mounting: English, as fitted by Sisson for equatorial sector. Solid polar axis 10 feet 3 inches long, telescope on one side of axis not counterpoised, 2-foot circles.

Driving-clock: none.

Apparatus: micrometer by Troughton.

Dome: hemispherical, 11 feet diameter.

Described by Airy as "an instrument of inferior class[10] ... a small telescope, mounted on a rude polar axis".[11] No illustration is known.

Historical Summary

c. 1827: Sisson's west sector converted into ordinary equatorial and 30-inch telescope fitted. 1828, 1832, 1835: single observations recorded in each year. 1846: dismounted by Simms. Polar axis and circles sent to Cape, telescope mounted on stand.

Contemporary Account

1845: Airy, *Astron. Obs.*, 1845.

8.3. *Sheepshanks 6·7-inch Refractor* (1838)

OG by Cauchoix of Paris, presented by Rev. R. Sheepshanks.

Mounting by F. Grubb, Dublin.

Mounted in S. Dome (later called Sheepshanks Dome), 1838–1963.

Mounted in Altazimuth Pavilion since 1963.

Still in use, 1975.

Description (when in Sheepshanks Dome: Fig. 81 shows it in the Altazimuth Pavilion.)

Telescope: 6·7-inch aperture, focal length 8 feet 2 inches, square wooden tube.

Mounting: German, using stone pier intended for Shuckburgh mounted as altazimuth (see p. 88). 12-inch hour circle divided to 1ᵐ, reading to 2ˢ by vernier. 11-inch declination circle divided to 15′, read to 30″.

Driving-clock: weight-driven with centrifugal governor.

Apparatus: double-image micrometer (from 1839). Comet eyepiece with thick wires and position circle (from 1890).

Astronomical Clock: Earnshaw.

Dome: hemispherical, 10·5-foot diameter.

Uses

Comets, occultations, double-stars (until about 1870), planetary measurement.

Historical summary

1837: OG by Cauchoix presented by the Rev. R. Sheepshanks, F.R.S. (1794–1855) Secretary of the Astronomical Society. Grubb given order for mounting. 1838: mounted in South Dome, built for Shuckburgh telescope east of Circle Room in 1813. First observations (Encke's comet) 29 October. 1839: double-image micrometer acquired. 1888: dismounted, cleaned and remounted. Adapted for stellar photography.[12] (It was seldom so used.) 1890: comet eyepiece acquired.

1914: OG used for eclipse of August 1914. 1949–50: some solar observations, after removal of photoheliograph to Herstmonceux. 1952: dismantled, reconditioned and re-erected in reconditioned dome. 1963: moved to Altazimuth Pavilion and handed over to National Maritime Museum. It remains operational. 1964: new finding telescope by Wildey.

Contemporary Accounts

1838: Telescope and mounting—Airy, *Astron. Obs.*, 1838., (i)-(iii). 1845: Double-image micrometer—Airy, *Astron. Obs.*, 1845, lxxxvi–xcii.

8.4. *Merz 12·8-inch Visual Refractor* (1859)
Airy's 'Great Equatorial'

OG by Merz of Munich, optical work by Troughton & Simms of London.

Mounting designed by G. B. Airy and made by Ransomes & Sims of Ipswich.

Water-driven clock by Dent.

Mounted in S.E. dome, 1860–91.

Mounted in Lassell dome, 1892–95.

As guiding telescope for 26-inch refractor in New Building from 1896 and at Herstmonceux (Dome E) from 1958.

Still in use, 1975.

Water clock preserved at National Maritime Museum.

Description (Figs. 82–83, 126)

Main telescope: 12·8-inch aperture, 17 feet 10 inches focal length visual refractor. Originally mahogany tube.

Finder telescope: 3-inch aperture, 2-foot focal length.

Mounting: English yoke type. Declination axis eccentric to polar axis to permit observations of celestial pole. 26-foot polar axis, framework of wrought iron. 24-foot cast-iron north pier weighing 5·5 tons. 5-foot declination circle, 6-foot hour circle.

Driving-clock: powered by water from Kent Water Company, using turbines (Barker's mill), controlled by mercurial conical pendulum and Siemen's chronometric governor.

Main apparatus: Airy double-image micrometer. 'Prismatic spectrum apparatus' (from 1863). Browning spectroscope (from 1874). Small camera (from 1875). Two built-in observing chairs.

Astronomical clocks: 'Arnold 2'. ½ sec clock controlled by sidereal standard. Chronometer Arnold 82, mounted near eyepiece, controlled by sidereal standard. Times of observations could be sent to chronograph by 'tapper' at eye-end.

Dome: wooden drum-shaped, 32-foot external diameter, shutters 4 feet width.

Method of use

Its original purpose was summed up by Edwin Dunkin in 1862:

> "This equatorial is provided with all the necessary adjuncts for making astronomical extra-meridional observations in every branch of the science. It is furnished with microscopes for reading the graduated circles, eyepieces of different powers and construction, and other appliances too numerous to mention. If the position of a planet or star be wanted: if the magnitude or diameter of Jupiter, Saturn, or any object having a disc, be required; or if the rapid changes noticed during an eclipse of the Sun are to be measured, or, indeed, the observation of any other phenomenon, this instrument, in the hands of a skilful observer, will give results which no one can doubt."[13]

Later, spectroscopic and double star observations were made.

Historical Summary

1845: existence of planet (later known as Neptune) outside orbit of Uranus predicted independently by Adams of Cambridge (September) and Le Verrier of Paris (November). Both predicted positions mathematically. 1846: new planet, searched for unsuccessfully by Challis at Cambridge from July 1846, detected by Galle in Berlin on 23 September, English public opinion considered this discovery ought to have been made in England. 1855: Airy presents Address to Visitors. After detailing deficiencies in present equatorials at Greenwich, Airy proposes new instrument whose design should be based on OG of 12 French inches by Merz, with mounting similar to Northumberland equatorial at Cambridge (designed by Airy himself) and water-power driving clock similar to that of Liverpool equatorial.[14] 1857: OG furnished by Merz. New southeast dome built—an octagonal three-storey building at east end. 1859: new telescope—the 'Great Equatorial'—erected. 24 May 1860: first recorded observation—occultation of Jupiter. 1862: experiments with spectroscope. 2 July 1874: first observation with new spectroscope by Browning. 1875: experiments with camera. 25 July 1877: ten-prism spectroscope replaced by half-prism spectroscope. 23 November 1891: 12·8-inch (Merz) refractor dismounted in preparation for placing 28-inch refractor on same mounting. May 1892: Merz telescope with Thompson photoheliograph attached mounted in Lassell dome south of Magnet House. 1893: observation of double stars in Lassell Dome. 1896: Merz telescope dismounted, given new tube, and remounted as guiding telescope to Thompson 26-inch photographic refractor. Eyepiece mounted on cross-slides for offset guiding.

1957: 26-inch refractor, with Merz OG as guider, re-erected at Herstmonceux (Dome E).

Contemporary Description

1868: General, *Astron. Obs.*, 1868, Appendix III. 1876: Spectroscope, *Astron. Obs.*, 1876, xxiii–iv.

8.5. *4-inch Dallmeyer Photoheliographs* (1873)

By T. R. Dallmeyer. One or other of five telescopes ordered for 1874 Transit of Venus. Used continuously at Greenwich from 1875 to 1949, then at Herstmonceux.

Still in use, 1975.

Description (Figs. 84–85)

An improved model of the Kew photoheliograph.

Of 4-inch aperture and 5-foot focal length, forming a primary image of the Sun a little over ½-inch diameter: this was enlarged by secondary magnifier to 4-inch diameter on 6-inch square camera plate. The length of the telescope was 8 feet. It was on a German equatorial mounting and had a driving-clock.

The wet plates first used were coated with a film of iodized cadmium collodion, and developed with pyro-gallic acid. A print from a 6-inch plate of 1883 can be seen in Fig. 85.

In 1884 a new secondary magnifier increased the size of the image to 8 inches diameter (strictly, 20 cm) on 10-inch square plates. The telescope length increased to 9 feet 7 inches.

Method of use (as fitted on Newbegin telescope from 1949)
1. Start driving-motor.
2. Set telescope on Sun by 'open sights'.
3. Centre Sun on ground-glass screen.
4. Adjust guiding telescope wires to limb.
5. Assess exposure. Set focal-plane shutter.
6. Remove ground-glass screen and insert carrier containing plate.
7. Check centring in finder.
8. Release shutter.

Historical Summary
1871: five identical photoheliographs ordered for British Transit of Venus expeditions of 1874 and 1882. 1874: all five erected for trial at Greenwich in their portable observatories. 21 July 1873: start of regular daily photographs of Sun at Greenwich by Dallmeyer. (Published observations start 17 April 1874.) April 1874: all Dallmeyers packed for shipment to Transit Expeditions. Kew photoheliograph started regular photos at Greenwich. September 1875: Dallmeyer No. 3 replaced Kew photoheliograph in wooden hut south of magnetic observatory. 1884: new secondary magnifier to increase image size to 8-inch diameter. 1891–4: new 9-inch Thompson photoheliograph came into use. Dallmeyer moved several times during building of New Physical Observatory. 1895–8: Dallmeyer on terrace roof of New Observatory. Thompson not used. 1898–1912: when Thompson photoheliograph was mounted on 26-inch refractor, Dallmeyer was only used when Thompson was sent on eclipse exhibition.

By 1910: Dallmeyer No. 2 replaced No. 3 on terrace roof as principal photoheliograph. New OG by Grubb. 1911: Dallmeyer No. 2 moved from roof of south wing of new building and mounted in Old Altazimuth Dome. New OG by Grubb. 12 February 1912: Dallmeyer used instead of Thompson as principal photoheliograph hereafter. 1929: spectrohelioscope observations started (see p. 120). 1939–45: photoheliograph observation continued during war. 2 May 1949: last photo of Sun taken at Greenwich. Second photograph taken same day by re-erected Dallmeyer, strapped to Newbegin 6¼-inch refractor at Herstmonceux in 22-foot dome where it is still (1974) in use.

8.6.　*13-inch Astrographic Refractor* (1890)
for the International Carte du Ciel

By Sir H. Grubb, F.R.S. of Dublin, 1888. New mounting by Grubb Parsons of Newcastle, 1969.

Mounted at Greenwich, 1890–1957 (except for occasional eclipse expeditions).

At Herstmonceux from 1958.

Still in use, 1975.

Description (Figs. 86–87)

Telescopes: 13-inch photographic refractor with parallel 10-inch visual guiding telescope, both of 11 feet 3 inches (343 cm) focal length (f = 10·4), 16-inch square photographic plates. Scale in focal plane: 1 mm to 1' of arc.

Mounting: German, with long declination axis to allow motion of 1½ hours beyond meridian on each side without reversing telescope. This and the double telescope weight necessitated a large counterpoise. New mounting by Grubb Parsons Ltd., 1969, very similar.

Driving-clock: weight-driven centrifugal governor placed inside stand, controlled electrically by seconds pendulum. Telescope provided with electric hand control by 1908.

Apparatus: duplex micrometer for plate measurement.

Astronomical clock: Dent 2017.

Domes: at Greenwich—18-foot hemispherical (papier-mâché on angle iron) with sectorial shutter opening 1/6 of the circumference of the horizon.

Dome at Herstmonceux: 22-foot hemispherical.

Method of use

Used for direct photography, originally for the international *Carte du Ciel*, for which two series of plates were taken, one of 4 ^{m} exposure on each area of 2° square for the chart, and another of three separate exposures of 6 ^{m}, 3 ^{m} and 20 ^{s} duration for the catalogue plates. From the catalogue plates, information was obtained on the positions and photographic magnitudes of stars; a repetition of the plates after an interval of time (20–30 years) gave the angular displacements or proper motions of the stars. Modern use of the telescope provides information on the colours of the stars also.

Historical Summary

April 1887: international *Carte du Ciel* project launched at International Astrophotographic Congress in Paris; all 18 participating observatories were to use similar telescopes of 343 cm focal length, giving a scale of 1′ of arc to 1 mm on the plate; Greenwich area of responsibility, a cap of 25° radius around the north celestial pole. August 1888: order for Greenwich telescope given to Sir Howard Grubb; new 18-foot dome erected over old Quadrant Room. May 1890: instrument delivered and installed, 30 feet above the ground, resting on the top of old Quadrant Pier of 1725. December 1891: regular series of photographs begun. December 1894: shutter blew off in gale; telescope was out of use until February 1895. January 1895: duplex micrometer brought into use for measuring the plates.

1903: eclipse expedition to Tunis; between eclipse expeditions the telescope was placed back, on its own mounting. 29 May 1919: OG used in Brazil to obtain photographs of total eclipse which proved validity of Einstein's theory of relativity. 1922: to Christmas Island for total eclipse. 1927: eclipse at Giggleswick, Yorkshire. 1929: eclipse in Siam. 1935: new controlling pendulum by Synchronome. 1939–45: out of use during World War II. 1947–8: employed on experimental work on photo-electric guiding for projected Isaac Newton 98-inch telescope. March 1950: pendulum ex-26-inch refractor fitted to drive. 1950–1: the only general purpose telescope available at Greenwich, used for observations of comets and the brighter minor planets. August 1957: installed at Herstmonceux (Dome D). 1969: new mounting by Grubb Parsons Ltd (see Fig. 87).

Contemporary Account

W. H. M. Christie, *Astrographic Catalogue 1900·0, Greenwich Section 1* (1904), i-xx.

8.7.   *Thompson 9-inch Photoheliograph* (1891)

By Sir Howard Grubb, F.R.S., of Dublin.

Presented 1891 by Sir Henry Thompson (1820–1904), distinguished surgeon and amateur astronomer.

Mounted at Greenwich, 1891–1939, except when on eclipse expeditions.

In store since 1939.

Description (Fig. 93)

Telescope (as at 1908): photograph refractor, 9-inch aperture, 8-foot 10-inch focal length, with Ross enlarging doublet of 4·3-inch focus. Sun's primary image 1-inch diameter, enlarged to 7·4 inches on photographic plate. Focal-plane shutter. In 1910 new camera fitted similar to that on Dallmeyer, with scale of 10 cm to solar radius (approx. 8-inch solar diameter).

Mounting: 1891–1939. Except when on eclipse expeditions, attached to tube of 26-inch refractor (p. 99) on the side opposite to the Merz guiding telescope (see Fig. 93).

Apparatus: 9-inch object-glass prism by Hilger.

Historical Summary

1891: presented by Sir Henry Thompson. Mounted on tube of Lassell reflector. Combined series with Dallmeyer. First photograph by Thompson, 17 December. April 1892: Lassell reflector replaced by Merz 12·8-inch in Lassell Dome. Thompson photoheliograph attached to Merz tube. 1894: building operations for New Physical Observatory prevented observations after 15 October. The telescopes were dismantled soon after. 1895: Japan for total eclipse. 26 May 1898: mounted on tube of 26-inch refractor in Thompson dome on top of new building. This was its permanent position until 1939, except when on eclipse expeditions.

1900: to Portugal for total eclipse. 1901: to Sumatra. 1905: to Tunis. 1910: new enlarging lens fitted. 1912: to Brazil. 1914: to Russia. Because of war conditions the Thompson photoheliograph remained at Pulkova observatory until 1925. 1925: returned from Russia. Mounted on Thompson equatorial. 1927: new enlarging lens by Ross. 1936: attached to tube of 30-inch reflector on other end of declination axis. 2 September 1939: dismounted on the outbreak of war. It has remained in store—at Greenwich and Herstmonceux— ever since.

8.8. *28-inch Visual Refractor* (1893)
The South-east Equatorial
The seventh largest refractor in the world

Telescope by Sir Howard Grubb of Dublin, 1893, on mounting by Ransomes & Sims of Ipswich originally used (1859–91) for Merz 12·8-inch refractor.

Mounted at Greenwich (South-east Equatorial building), 1893–1947.

Mounted at Herstmonceux (Dome F), 1957–71.

Transferred to National Maritime Museum and re-erected at Greenwich, 1971.

Description (Figs. 88–91)

Main telescope: 28-inch aperture, 27-foot 10-inch focal length, photo visual refractor. Though seldom done, the crown-glass component of the object glass could be reversed and the separation increased to alter the correction for chromatic aberration from that appropriate for a lens for visual use to that suitable for photographic use with blue-sensitive plates (focal length reduced by 23 inches).

Finder/guiding telescope: 6½-inch aperture, 8-foot focal length, from Corbett equatorial (purchased for the use in Transit of Venus expedition 1875).

Mounting: English equatorial, as for Merz 12·8-inch refractor already described.

Driving-clock: water-powered (as described under Merz 12·8-inch refractor), 1893–1929. Electric drive of the Gerrish type, controlled by pendulum, 1929–47. Synchronous motor, 1957–71.

Astronomical clock: Dent 2009 until 1929, then Molyneux with wooden pendulum.

Apparatus: position and transit micrometers and eyepieces of Merz 12·8-inch refractor adapted to new telescope. Double-image micrometer.

Dome: at Greenwich, onion-shaped, with 36-foot diameter contracting onto 31-foot base. This was necessary because the 28-inch telescope was 10 feet longer than the 12·8-inch telescope it replaced on the same mounting. Shutter opening 7 feet wide extending from horizon to horizon. At Herstmonceux, 37-foot hemispherical.

Method of use

A few visual spectroscopic observations in early days, thereafter used almost exclusively for double-star observations.

Historical Summary

1885: decision taken that Merz 12·8-inch refractor should be replaced on the same mounting by more powerful telescope. Order given to Grubb. 23 November 1801: Merz telescope dismounted from S.E. Equatorial dome. November 1892: old wooden drum dome dismounted. 16 December 1892: 25 April 1893: new 36-foot 'onion' dome erected by Messrs T. Cooke & Sons of York (Figs. 88 and 89). January–June 1894: trials of photographic observations. September 1894: regular double-star observations started. September 1898: balcony erected around SE equatorial dome.

1914–18: regular observations suspended due to war conditions. October 1919: instrument dismantled for replacement of upper pivots worn by continuous use for 68 years. October 1920: repairs completed. 1929: Gerrish electric drive replaced water clock which was presented to Science Museum, London (later transferred to NMM). Early 1939: dome rail fractured so that dome could not be moved. September 1939: 28-inch OG dismounted and sent to a place of safety. October 1940: papier-mâché covering to dome holed in air raids. 15 July 1944: fall of VI bomb in park not far from 28-inch dome caused it to be stripped of most of its covering. Mounting undamaged. Autumn 1947: telescope and mounting dismantled and sent by road to Herstmonceux "The north pier, a single 5-ton casting 22 feet high, which must be a unique specimen, was successfully run out through a breach in the dome wall, down a ramp and on to a lorry"[1]. June 1953: dome at Greenwich removed and replaced by flat circular covering. July 1957: telescope and mounting erected in Dome F at Herstmonceux. September 1957: telescope brought into use at Herstmonceux; trials carried out on photographic use, but at visual focus with panchromatic plates and yellow filter, not as described above. 11 July 1958: new double-star programme started. October–November 1971: telescope and mounting dismantled at Herstmonceux, transported by road to Greenwich, and re-erected on South-east Equatorial building, using original 1857 holding-down bolts. April 1974: work started on new Onion dome at Greenwich. The appearance and operation are the same as the 1893 dome but modern materials are used.

Contemporary Accounts

1908: Christie, *Astron. Obs.*, 1908, xviii. 1921: F. W. Dyson, *Catalogue of Double Stars*, 1921, iii–vi.

8.9.   *Thompson 26-inch Photographic Refractor* (1896)

Signed: *Sir Howard Grubb, Dublin, 1896.*

Gift of Sir Henry Thompson (1820–1904), distinguished surgeon and amateur astronomer.

Mounted at Greenwich with Thompson 30-inch reflector on same mounting, 1897–1947.

Mounted at Herstmonceux (Dome E), with counterweight instead of 30-inch reflector, 1957.

Still in use 1975.

Description (Figs. 92–94)

Main telescope: 26-inch aperture, 22-foot 5-inch focal length, photographic refractor. Scale at focal plane: 30″·2 to 1 mm on photographic plate—double that of 13-inch astrographic refractor.

Guiding telescope: Merz 12·8-inch refractor already described fitted with new steel tube.

Mounting: German equatorial, modified to allow of complete circumpolar motion without reversal with (at Greenwich) Thompson 26-inch refractor, Merz 12·8-inch refractor (as guiding telescope) and Thompson 9-inch photoheliograph at one end of declination axis; Thompson 30-inch reflector and Hodgson 6-inch refractor (as guiding telescope) at the other.

Driving-clock: originally, Grubb drive, in base of mounting, controlled by clock Dent 2016. From 1936, synchronous electric motor.

Astronomical clocks (1905): Dent 2016 with seconds contacts; Graham 2 (for longitude determination).

Domes: at Greenwich, 30-foot hemispherical, papier-mâché on angle iron. Single curved shutter opening 3 feet 6 inches at zenith, 6 feet at horizon, originally made by T. Cooke & Sons of York for Lassell telescope (*q.v.*) in 1884, placed on central tower of New Observatory in 1896. At Hertsmonceux (Dome E), 34-foot hemispherical, with rising floor.

Method of use (as at 1974)

Used for photographic parallax measurement, proper motions and photometry.

1.   Prepare telescope (checks, focus, etc.).
2.   Set sidereal time on driven RA circle.

3. Set telescope to RA and Dec of object. Locate object in finder and centre on crosswires, or in autoguider—a photoelectric device for keeping star centred.
4. Insert plate carrier, loaded with appropriate type of 16 cm square photographic plate, into breech-end which provides positive location of the telescope focus.
5. If guiding visually, correct with manual remote control. More usually, use autoguider.
6. In parallax work, brightness of parallax star is reduced by means of an appropriate occulting shutter to produce an image of similar intensity to that of the comparison stars. Exposures of the order of 5 minutes are typical of this work. For greater accuracy, the plate is rotated through 180 degrees and another exposure given.
7. The telescope is also used for monitoring the light variations of quasars, which, because of the faintness of these objects, necessitates long exposures.

Historical Summary

1894: Sir Henry Thompson offered the sum of £5000 to provide a large photographic telescope to complement the 28-inch visual telescope just completed; Merz 12·8-inch refractor, which was to have been main telescope on central tower of New Observatory, to become guiding telescope; Thompson 9-inch photoheliograph to be placed on same mounting. 5 May 1894: order placed with Sir Howard Grubb. 1896: Thompson presented new 30-inch photographic reflector (*q.v.*) to go as counterpoise on other end of declination axis. September 1896: Lassell dome erected on central tower of New Observatory. April 1897: telescope brought into use.

1936: extensive alterations to alleviate cramped conditions. 2-foot section removed from upper part of tube and corresponding section added at breech end: sliding dew cap installed; driving clock replaced by synchronous motor; Thompson photoheliograph attached to tube of 30-inch reflector instead of 26-inch refractor. 2 September 1939: 26-inch OG dismounted and removed to a place of safety. Autumn 1947: telescope and mounting dismantled and sent to Grubb Parsons at Newcastle to have roller bearings inserted in both axes. 1957–8: telescope and mounting re-erected at Herstmonceux in Dome E (with rising floor). Counterpoise replaces 30-inch reflector at one end of declination axis.

Contemporary Account
1905: Christie, *Astron. Obs.*, 1905, xviii–xx.

8.10. *Thompson 30-inch Photographic Reflector* (1896)

By Sir Howard Grubb of Dublin, under supervision of Dr Common.

Gift of Sir Henry Thompson, 1896.

Mounted at Greenwich in Thompson equatorial, 1897–1947.

Mounted at Herstmonceux on fork mounting by Cox, Hargreaves & Thomson, 1956.

Still in use, 1975.

Description (Figs. 95–96)

Main telescope: 30-inch Cassegrain reflector. Focal length of primary 11 feet 5 inches (f:4·5), 1 mm on photographic plate at prime focus corresponding to a fraction less than 1′ of arc. Two convex mirrors of 24-foot and 38-inch focal length give equivalent focal lengths at the Cassegrain focus of 76 feet and 49 feet respectively. Coudé system (f:47) added 1963.

Guiding telescope: Hodgson 6-inch refractor, OG said to be by Cauchoix,[15] 6-foot 6-inch focal length, ex Transit of Venus (Hawaii), 1874, purchased from R. Hodgson, originally on long polar axis similar to Lee equatorial (see p. 118).

Mounting: at Greenwich, see Thompson equatorial already described. At Herstmonceux, fork mounting by Cox, Hargreaves & Thomson.

Apparatus: slitless spectroscope made in workshops.

Dome: at Greenwich, see p. 99 above. At Herstmonceux (Dome A), 28-foot hemispherical.

Method of use

Stellar spectroscopy, with photographic or image-intensifier recording in recent years.

Historical Summary

1896: presented by Sir Henry Thompson. Dr Common figured mirrors and supervised construction. 1897: erected on Thompson equatorial mounting at opposite end of declination axis to 26-inch refractor. 1898: brought into use.

February 1908: Melotte discovered eighth satellite of Jupiter with photographs from this telescope. 1936: Thompson 9-inch photo-heliograph mounted on tube. September 1939: mirror removed to a place of safety. 1947: Thompson equatorial dismounted at Greenwich. December 1956: mounted on new fork-mounting in Dome A at Herstmonceux. Mirror re-figured by Cox, Hargreaves & Thomson.

Contemporary Acccount
1905: Christie, *Astron. Obs.*, 1905, xviii–xx.

8.11. *Yapp 36-inch Reflector* (1932)

Signed: *Grubb-Parsons, Newcastle-on-Tyne, 1932.*
Gift of William Johnston Yapp, industrialist, 1931.
Mounted at Greenwich, 1934–55.
Mounted at Herstmonceux, 1958.
Still in use, 1975.

Description (Figs. 97–98)

Main telescope: 36-inch Cassegrain reflector on glass, 15-foot focal length (f:5), 11-inch convex quartz secondary giving equivalent focal length of 45 feet for slit spectrograph. 7-inch convex quartz secondary to give parallel beam for slitless spectrograph. Open lattice-work tube, later partly closed in.

Two finder-guiding telescopes: 7-inch and 3-inch refractors.

Mounting: modified English equatorial with 24-foot solid cast-iron polar axis. Cross-head attached to middle of axis carries telescope on one end and counterpoise on the other.

Driving-clock: modified Grubb drive.

Apparatus: slitless spectroscope by Hilger (1933) (no longer used). One prism/three prism slit spectroscope by Hilger (1937) (no longer used). Currently used with image-intensifier spectrograph and photometer, designed and built by RGO.

Dome: Greenwich, 2-storied brick building in Christie enclosure. 34-feet external diameter hemispherical dome with electrically-operated 7 feet 6 inches opening. Copper covered. Herstmonceux (Dome B), 34 feet hemispherical.

Method of use

Used with old spectroscopes for measurement of the colour temperatures of stars. Now used for photo-electric photometry, and stellar spectroscopy with image-intensifier spectrograph.

New spectrograph has exposure meter.

Spectra obtained on plates or film on which an arc spectrum is exposed *in situ* with a Hartmann diaphragm.

Historical Summary
March 1931: Mr W. J. Yapp gave £15,000 for purchase of 36-inch reflector, slitless spectroscope and 34-foot dome. Order placed with Grubb Parsons for telescope and dome, and with Hilger for spectroscope. November 1932: new dome completed in former Magnetic Observatory (Christie enclosure) in Greenwich Park. 20 April 1934: telescope in operation. April 1937: slit spectroscope by Hilger delivered. 2 September 1939: mirrors dismounted and sent to a place of safety. January 1946: mirrors sent for aluminizing. December 1946: observations restarted at Greenwich. March 1948: observations ceased because of poor condition of aluminized mirror, caused by pollution of Greenwich atmosphere. Mirror silvered and brought back into operation. November 1955: telescope dismantled and sent to Herstmonceux. 1957: mirror re-figured by Grubb Parsons. August 1958: brought into use in Dome B at Herstmonceux.

Contemporary Account
1934: Sir Howard Grubb, Parsons & Co. 'The 36-inch Greenwich Reflector', *The Engineer*, May 1934, 500–2, 524–7 and supplement 1934: Spencer Jones *Report* 1934, 3–4.

8.12. *98-inch Reflector* (1967)
The Isaac Newton Telescope
Signed: *Grubb–Parsons England 1967.*
Mounted at Herstmonceux, 1967.
Still in use, 1975.

Purpose
For direct stellar photography, photographic and electronographic spectrography, photometry and interferometry.

Description (Figs. 99 and 100)
Main telescope: Primary mirror 98-inch (250-cm) aperture, 24-foot 7-inch (750-cm) focal length, supported on pressurized airbag. Available foci: prime (f:3·3 with Wynne coma-correcting lens), Cassegrain (f:14), Coudé (f:32).

Two guiding telescopes: 20-cm refractor, 300-cm focal length; 40-cm Cassegrain reflector, 650 cm focal length, fitted with low-light-level TV camera.

Mounting: fork-type, with polar axis on oil-pressure-supported polar disk.

Apparatus: for direct photography at prime and Cassegrain foci; image-intensifier spectrograph, photographic spectrograph, photometers for use at Cassegrain focus; Coudé spectrograph. Visiting astronomers frequently use their own equipment.

Dome: 60-foot hemispherical.

There is a separate control console. The observer is located either at prime focus or in chair at Cassegrain focus (as Fig. 100).

Historical Summary

1949: Pyrex disk presented by McGregor Trust, Michigan, U.S.A. 4 December 1959: order for building of telescope and dome placed with Grubb, Parsons, Ltd. Summer 1967: commissioned. 1 December 1967: inaugurated by H.M. The Queen.

Contemporary Description

1967: P. Lancaster Brown, 'The 98-inch Isaac Newton Telescope', *Sky and Telescope* (December 1967), 356–61.

8.13.   *38-inch Hargreaves Reflector* (1960)

By Cox, Hargreaves & Thomson, 1960, originally intended for Elisabethville, Katanga. Purchased by RGO 1972.

Still under test, 1974.

Description

Main telescope: primary mirror 96-cm aperture, 193-cm focal length. Available foci: prime (f:2), Schmidt camera (f:3), Cassegrain (f:10,) Coudé (relayed Cassegrain f:10).

Guiding telescope: OG 25-cm aperture, 378-cm focus, by Cave. 1973.

Mounting: single pier with counterweight.

Dome: 37-foot hemisphercial (Dome F).

NOTES

1. For Sharp, see p. 21. The telescope was described by Sharp in a letter to Flamsteed dated 17 February 1702, quoted by Cudworth, pp. 63–4. See also *MNRAS*, III (1835), pp. 117–8 and *The Observatory*, 772 (1938), pp. 248–50.

2. O. Gingerich, "What is an English mounting", *Sky and Telescope*, 34 (November 1967), 5.

3. G. B. Airy, *Account of the Northumberland Equatorial and Dome*, (Cambridge, 1844), 3.

4. R. Hooke, *Animadversions . . .* (1674), 69.

5. W. H. Smyth, *The Cycle of Celestial Objects continued.* . . . (*Speculum Hartwellianum*), (1860), 158.

6. *Phil. Trans.*, 1793, 76n.

7. Pearson, p. 518.

8. G. B. Airy, *Address to the . . . Board of Visitors . . .*, Greenwich, 18 October 1858.

9. Airy, *Astron. Obs.*, 1862, Appendix II (15).

10. Airy, *Astron. Obs.*, 1845.

11. Airy, *Address to Visitors 1855* [3].

12. Christie, *Report*, 1888/10.

13. E. Dunkin, "A Day at the Observatory", *Leisure Hour*, 1862, 39.

14. G. B. Airy, *Address . . . by the Astronomer Royal*, Greenwich (1855).

15. G. B. Airy (Ed.), *Transit of Venus 1874 . . .* (1881), 222.

9

The Minor Astronomical Instruments

Flamsteed's Refracting Telescopes

WHEN Flamsteed came to Greenwich in 1675, he brought with him two refracting telescopes, one of 8½-feet, another of 16-feet focus, together with the famous Towneley micrometer to be described later. These he used, generally in the Great Room, for observations requiring to be timed by clock—eclipses, occultations, etc.—and those demanding measurement of small angles by micrometer—Moon and planet diameters, appulses (close approaches) and the like. These telescopes, illustrated in the Great Room in Fig. 2 and on the roof in Figs. 4 and 68, had square wooden tubes, apertures of about 2 inches, and provision for fitting a micrometer (Fig. 101).

The 8½-foot tube was referred to in observation books as *tubus breviore*, *tubus minore* or *tubus 8 ped*, the 16-foot tube as *tubus longiori* or *tubus ped. 16*. We know their precise focal lengths—8 feet 7½ inches and 16 feet 4½ inches[1]—but we do not know the apertures or makers, though Coxe and Yarwell are often mentioned in correspondence.

About 1682, Flamsteed acquired a 27-foot telescope (until about 1820 telescopes were referred to by their focal lengths rather than by their apertures). He used this larger telescope almost exclusively for timing eclipses of Jupiter's satellites.

The following numbers of observations were recorded:

Teles-cope	1676–85	1686–95	1696–1705	1706–15	1713–19	Total
8½ ft.	61	19	5	6	2	93
16 ft.	150	95	29	2	—	276
27 ft.	4	21	15	—	—	40
Total	215	135	49	8	2	409

In addition, the 7-foot telescopes attached to the sextant were some-times used as additional 'gazing' telescopes for important eclipses, etc., particularly when there were visiting observers.

Flamsteed's Micrometers

When Flamsteed first met Jonas Moore in 1671, he was presented with the Towneley micrometer, apparently the very instrument demonstrated to the Royal Society in 1667—Richard Towneley's improved version of that of the inventor William Gascoigne (1616?–45) whose original instrument had been acquired by his uncle Christopher Towneley after Gascoigne's death in the Civil War.[1a] Designed to measure angular distances between objects in the telescope field, the Towneley instrument is illustrated in Fig. 102.

Bringing it with him when he came to Greenwich in 1675, Flamsteed continued to use it with his 16-foot and 8½-foot telescopes until just before he died. His observation books recorded both 'revolves of the screw' and the equivalent angle in arc-minutes, from which the following table is derived:

Revolves of the Screw	1000	2000	3000	4000	5000	6000	7000
8½ foot telescope	8'	16½'	25'	33'	41'	54'	—
16 foot telescope	4½'	9½'	14½'	19½'	25'	29½'	35'

We know that Flamsteed had at least two other micrometers—one "made by one Whitehead",[2] another to the design of Abraham Sharpe given to him in August 1704.[3] However, all the published micrometer measurements from 1676 to 1718 seem to conform to the table shown above, indicating either that the Towneley micrometer was used exclusively or that the screw of the Whitehead instrument had the same number of threads to the inch.

The 2-foot Solar Telescope

Francis Place's etching *Domus obscurata* (Fig. 102), probably executed in 1676, shows Flamsteed's solar observatory in the north-east summer house. Documentary references occur only with the eclipse of 2 July 1684, however. Then Flamsteed described his observing method as follows, explaining elsewhere that he was following the practice of Gassendi and Hevelius:[4]

> "I observed the eclipse of the Sun both through a telescope of 16 ft and also on a scene [screen] in a darkened room. The diameter of the image transmitted through the 2 ft. telescope above the scene was 5½ English inches, which was divided into 16 concentric equidistant circles of 32 minutes, as in the diameter of the Sun".[5]

The Sirius Telescope

In connection with Flamsteed's investigations into the rotation of the Earth (see p. 125-6) Flamsteed needed to measure the length of the sidereal day. His method is explained in a letter to Towneley in 1677:

> "I have now fixed a pair of convex glasses of six foot focus in brass cells upon an iron ruler to one of our walls [on a balcony of the Great Room] and 3 or 4 times observed the transits of Sirius over it."[6]

The telescope was fixed at an altitude corresponding to the declination of Sirius which, being the brightest star in the sky and lying some 40° from the Sun's annual path, can be seen in a telescope night and day the year round.

The Mast Telescope

Before the invention of the achromatic telescope, the only way to reduce the deleterious effects of chromatic and spherical aberration in a refractor was to reduce the curvature of the objective—but this increased the optical focal length and thereby the physical length of the telescope.

A long telescope of this sort was a striking feature in early pictures of the Royal Observatory (Fig. 5, for example) which shows an

80-foot mast in the garden, from which is suspended a 60-foot telescope tube.

Pepys reports that the Navy Board supplied Jonas Moore with two or three unserviceable masts for the observatory in March 1676. Hooke's diary for 1 May says: "Sir Jonas Moore erected pole at Greenwich. All the conjurors there". As the 80-foot mast had to remain upright without stays, this must have been quite an operation. One would like to know just who the 'conjurors' were.

Through the Royal Society in June 1677, Moore acquired an object-glass of about 60-feet focus made by 'the Parisian'—presumably Pierre Borel who made the 87-foot OG for the well telescope (p. 58).[8] On 24 October 1677 an observation of Jupiter's satellites was recorded (the only one), the telescope being described as of 58-foot focus in the MS observation book,[9] of 57 feet in the printed version.[10]

In November, Flamsteed reported:

> "I have got the glass of 60 ft tried, but find neither it nor Mr. Coxe's of 48 ft to be any ways excellent. They show Jupiter large enough but I think I see his belts distincter in a 16 ft. glass than in either. I am confident by those observations which Mr. Cassigny lately imparted that the French glasses excel ours far. How we shall get any better, I know not."[11]

We hear nothing further of the mast telescope—which obviously proved to be no success—until 12 July 1690 when Flamsteed reported that the previous night the mast "began spontaneously (since there was no wind blowing) to sway to and fro, and with its top shuddering, it threatened to crash down".[12] Just as a woodcutter had been summoned to fell the pole, the swaying ceased and so they "fortified it against the might of the wind with handsome supports". This event caused Flamsteed concern in case it should have crashed down on the Great Room of the Arc House.

There are no more documentary references but, by the time Gasselin sketched the Observatory in 1699, the mast had been removed.

Flamsteed's 3-foot Quadrant

When he was appointed the King's observator, Flamsteed brought with him to Greenwich a quadrant of 3-foot radius, probably of wood, with a brass limb divided by himself and with a quicksilver level—evidently the instrument seen in Fig. 2. It was capable of rotation about a vertical axis, 'voluble' being the contemporary description. This he used primarily for time-determination by equal-altitude observations—by noting the clock-time the Sun or a star was at a certain altitude before culmination, and again at precisely the same altitude after culmination: halfway between the two gave the moment of meridian passage—apparent noon for the Sun.

The Hooke Screwed Quadrant

In January 1677, against the opposition of Robert Hooke, Sir Jonas Moore persuaded the Council of the Royal Society to lend the Observatory their 3-foot quadrant—made in 1674 by Tompion to Hooke's design, which was lying idle.

As described and illustrated by Hooke in his *Animadversions . . .* of 1674, this instrument was of extraordinary ingenuity, some features being 150 years ahead of their time:[13]

(*a*) the eyepieces for both the fixed and the moving telescopic sights were placed one above the other at the centre of the quadrant instead of on the circumference as was usual: with the aid of two diagonal mirrors both images could be seen simultaneously by one observer—rather as in a marine reflecting sextant, though differently contrived;

(*b*) the position of the moving telescope on the quadrant was read exactly as in a modern marine micrometer sextant—and as in Flamsteed's 7-foot sextant—by revolves of the perpetual screw rather than by diagonal or vernier scales;

(*c*) angles of up to 180° could be measured by making the fixed telescope (which had two object glasses) capable of looking either one way or the other by altering the setting of the diagonal mirror;

(*d*) a water level gave the perpendicular for altitude observations;

(*e*) a 'German' equatorial mounting with counterpoise, described but never fitted;

(*f*) a driving-clock, regulated by a conical pendulum and driven by a weight, was also described but not fitted.

Flamsteed needed it primarily for time-determination so he had it mounted on an iron vertical axis. Having had poor results on his sextant by measuring angles by 'revolves of the screw', he set about placing a diagonal scale on the brass limb. He first used it in the Great Room on 16 June 1678—for time-determination and later for measuring refractions.[14]

On 27 August 1679, Moore died. Hooke determined to take his revenge: less than a month later—on 22 September—the Royal Society Council ordered that all its instruments should be removed from Greenwich as soon as possible. Wren, Hodgkins and Hooke collected Hooke's quadrant and a few other small items on 26 September, refusing to allow Flamsteed even a few days to make alternative arrangements.

". . . It was so ill-contrived . . . that I could not make it perform better than my first."[16] So Flamsteed described the Hooke quadrant later. Does one detect a note of sour grapes?

Flamsteed's 50-inch Voluble Quadrant
Flamsteed continued:

> "And now he obliged me to think of fitting up one of my own, of 50 inches radius, wherein by peculiar contrivances I had avoided all the inconveniences I had met with in his. This gives an observed height to half a minute: and now, by it, I am sure of the observed time to three seconds; which I could not have expected from either of my other instruments."[17]

Unfortunately, we do not know exactly what the 'peculiar contrivances' were; only that the quadrant was of iron and had a radius of 50 inches; that the diagonals were drawn with the point of a sharp needle and that the altitude could be read to $\frac{1}{8}$ minute easily; that the plumb-line was of fine silver wire passed through a candle and that the plummet weighed something more than 1 oz.[18] His previous altitude-measuring instruments had the quadrant itself fixed in the horizontal by reference to some sort of bubble level, with the telescopic sight moving on the limb, as with the mural arc. In this latest one, it seems likely the sight was fixed to the frame and that, to take an observation, the whole quadrant was rotated in the vertical plane, the altitude being read off on the limb against a plumb-line suspended from the centre. Such a quadrant to Flamsteed's design, sent to William Molyneux in Dublin in 1682, is illustrated in Fig. 103.[19]

The first recorded observation was on 20 October 1679, three weeks after the Hooke quadrant was so suddenly removed. After Flamsteed's death, there was talk of Molyneux (probably Samuel not William) offering to buy it and then refusing.[20] It has not been heard of since.

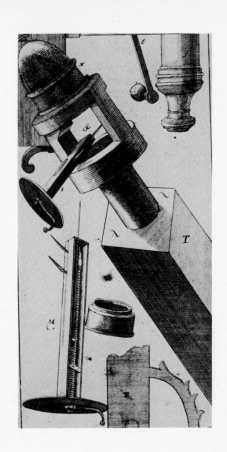

Fig. 101. THE TOWNELEY MICROMETER ABOUT 1676

y. T The tube of a refracting telescope. *x* The telescope eyepiece.
Where the eyepiece fits into the main square tube. The circular tube
could be pushed in and out to allow the heavenly bodies and the
micrometer sights to be brought into focus simultaneously, and it
could be rotated within the square tube to get the micrometer at the
required inclination. *œ* The aperture into which the micrometer is
fitted when required. (The telescope was often used without the
micrometer.) When the micrometer was in place, there was a gap
through which light shone to illuminate the sights. At night, a
lantern would be used. *i* The micrometer in its working position in
the eyepiece. *μ* A detailed view of the micrometer. The distance apart
of the two sights is controlled by a handle at the bottom, and is read
off (in terms of threads or 'parts') as follows:
whole numbers—on a scale on the bar protruding from the top of
the drawing; hundredths—on the circular scale at the bottom of
the drawing. The 'parts' were converted into minutes and seconds
of arc by reference to a table.
Detail from the etching *Partes Instrumentorum pertinentium ad
Speculam Astronomicam*, by Francis Place after Robert Thacker,
1676.

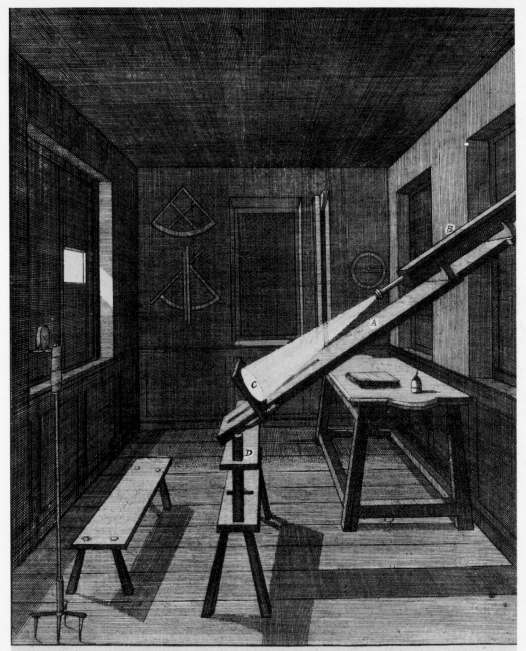

Domus Obscurata, ad Maculas, Eclipsesque Solares Excipiendas, peropportuna.

Fig. 102. FLAMSTEED'S SOLAR OBSERVATORY, ABOUT
1676
"Darkened house, very convenient for taking sunspots
and eclipses", the summer house at the east end of the
terrace. *Key. A* Support for the telescope and screen.
B 2-foot refracting telescope projecting Sun's image.
C Screen on which 5½-inch diameter image of Sun was
projected. *D* stand for *A*, adjustable in height. From
the etching *Domus obscurata. . .* by Francis Place after
Robert Thacker, *c.* 1676.

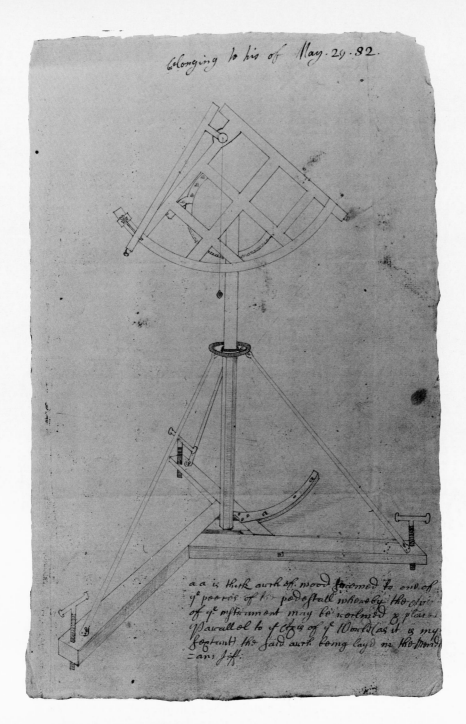

Fig. 103. A 'VOLUBLE' QUADRANT 1682

probably similar to Flamsteed's 50-inch quadrant of 1679. The quadrant illustrated was designed by Flamsteed for William Molyneux of Dublin in 1682. The text reads:

> *aa* is [a] thick arch of wood screwed to one of the pieces of the pedestal whereby the axis of the instrument may be reclined and placed parallel to the axis of the world (as it is [with] my sextant) the said arch being laid in the meridian :J:F:

From a letter from J. Flamsteed to W. Molyneux, 29 May 1682 (D/M. 1/3, f. 24). By permission of the Southampton City Record Office.

Fig. 104. JAMES SHORT, F.R.S., OPTICIAN (1710–68)
who made several reflecting telescopes used at Greenwich. One of
the Commissioners for the Longitude, he was candidate for the post
of Astronomer Royal in 1765. From an oil painting at the Royal
Observatory, Edinburgh. By permission of the Astronomer Royal
for Scotland.

Fig. 105. JOHN CHARNOCK'S SKETCHES IN THE GREAT ROOM
ABOUT 1785
Top left 5-foot achromatic refractor by Dollond *Top right* John
Harrison's second marine timekeeper *Bottom* 6-foot Newtonian reflec-
tor by James Short. From 'Charnock's Views', Vol. iv., in NMM
Print Room.

Fig. 106. PETER DOLLOND, F.R.S., INSTRUMENT MAKER
AND OPTICIAN, (1730–1820)

Eldest son of John Dollond, inventor of the achromatic
telescope, he was in partnership with his father until
the latter's death in 1761. Peter was an astute business
man. Having secured the patent for the achromatic lens,
he made certain that, by 1770, no serious telescope user
anywhere in Europe would continue to use a simple
object-glass. Between 1765 and 1820 he fitted achromatic object-glasses to all the major instruments at Greenwich and supplied many smaller telescopes. From an
oil painting by an unknown artist in possession of
Dollond & Aitchison Ltd.

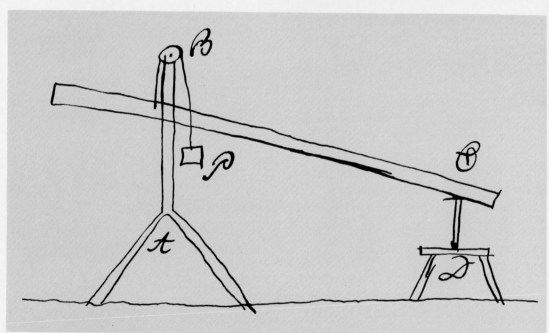

Fig. 107. 10-FOOT ACHROMATIC TELESCOPE BY DOLLOND

sketched by a visiting Danish astronomer, Thomas
Bugge in 1777. In his diary, Bugge says:

> "The 10-foot Dollond was mounted as shown in the figure.
> The front part is hung from a pulley B attached to the
> support AB and counterbalanced by means of the weight
> P. The back part C is supported by a footing D whose

bar is threaded. It can be raised or lowered in the same
way as is shown in my description of the astronomical
chair."

From Bugge's diary, p. 87 right by permission of the
Royal Library, Copenhagen (N.y. K.g.l. saml. 377c,
4to).

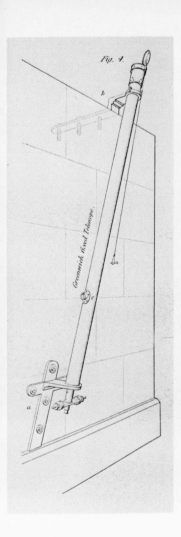

Fig. 108. ALPHA CYGNI 10-FOOT FIXED TELESCOPE, 1816
mounted on the west side of the Quadrant pier (where the bolt holes
can still be seen) from 1816 to 1839. Dr Brinkley of Dublin claimed
to have found a large parallax for the star α Cygni. In 1816, Pond
mounted this telescope at a ZD of 6° 50′—on the star's declination—
and found no sensible parallax. From W. Pearson, *Practical
Astronomy*, II, Plate XII, Fig. 4.

Fig. 109. PORTABLE ALTAZIMUTH WITH AXIS VIEW
made by Troughton & Simms in 1866 and used in New Zealand
for the Transit of Venus in 1874. The eyepiece is to the left of the
picture, at the end of the axis: the 15-inch vertical circles were read
by microscopes in the mounting (which cannot now be found).

Fig. 110. PORTABLE ALTAZIMUTH,
one of four made by Troughton & Simms specially
for the Transit of Venus expedition of 1874 and 1882.
The vertical circles were of 14-inch diameter.

Fig. 111. A PORTABLE TRANSIT OBSERVATORY FOR THE 1874
TRANSIT OF VENUS EXPEDITION TO HAWAII;
showing clock Dent No: 1916 and the $3 \times 36\frac{1}{2}$-inch transit instru-
ment later used for time-determination at Greenwich and
Abinger. From *Stargazing: Past and Present* by J. Norman Lockyer,
FRS (1878), p. 236.

Fig. 112. THE MAGNET HOUSE ABOUT 1870,
looking north-east. The 79-foot mast carried the electrometer whose
lamp is being cleaned. The thermometer stand can be seen in front.
From a drawing in possession of The Royal Greenwich Observatory.

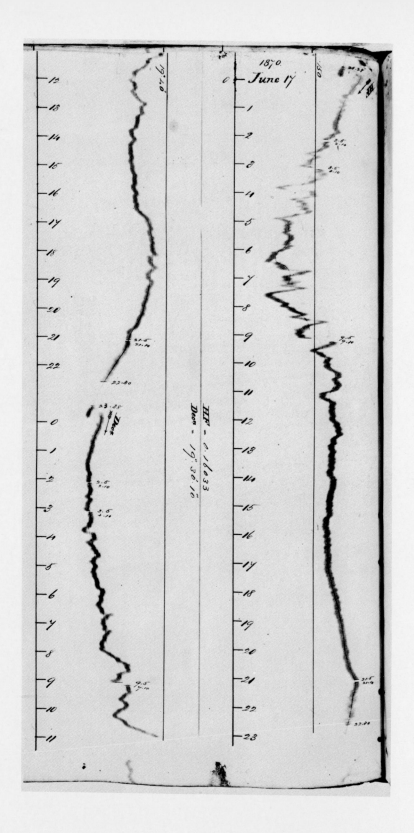

Fig. 113. PHOTOGRAPHIC REGISTRATION OF MAGNETIC DECLIN-
ATION AND HORIZONTAL FORCE,
started in 1847.

Fig. 114. THE MAGNETIC PAVILION IN THE CHRISTIE ENCLOSURE
erected 1898 for instruments measuring absolute magnetic elements.
From a photograph by Lacey Maunder.

Fig. 115. AIRY'S DIP CIRCLE,
by Troughton & Simms, used from 1861 to 1914.

Fig. 116. DIP INDUCTOR

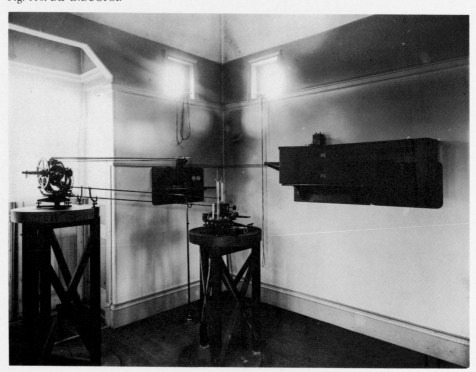

Fig. 117. **KEW PATTERN UNIFILAR MAGNETOMETER**

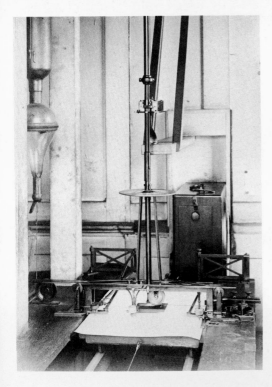

Fig. 118. **THE OSLER ANEMOMETER**
in the NW Turret on Flamsteed House, from 1840 to
1953. The Pluviometer (recording rain gauge) is on the
left.

Fig. 119. ROBINSON ANEMOMETER,
on a special platform on the roof of Flamsteed House.

Fig. 120. MEASURING THE WIND, 1881
From *The Illustrated London News*, 31 January 1880, p. 101.

Fig. 121. TAKING TEMPERATURES, 1881
From *The Illustrated London News*, 31 January 1880, p. 101.

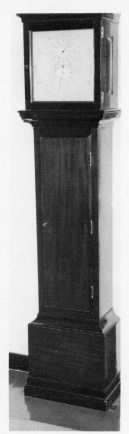

Fig. 122. TOMPION YEAR-CLOCK: MOVEMENT WITH PENDULUM ABOVE
This is a photograph of the replica installed in the Great Room (Octagon Room) in 1958. Note the upward-facing crutch embracing the tail of the 13-foot pendulum.

Fig. 123. 'GRAHAM 3', TRANSIT CLOCK AT GREEN-WICH FROM 1750 TO 1821,
seen also in Fig. 32.

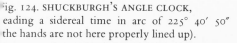

Fig. 124. SHUCKBURGH'S ANGLE CLOCK,
reading a sidereal time in arc of 225° 40′ 50″
(the hands are not here properly lined up).

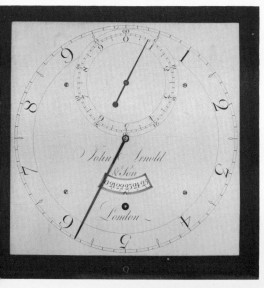

Fig. 125. TAYLOR'S ALARM CLOCK,
signed: *Johnson Strand London* 8-day, keeping sidereal time, with anchor escapement; pull-wind alarm (ringing bell) with silencing lever, set by inserting pins any 10^m. Strikes gong at the hour and 3^m before the time of transit of each 'clock star', as named on dial.

Fig. 126. THE OBSERVATORY IN 1885
(*Top, left to right*) (1) Airy Transit circle, (2) Great Equatorial, (3) Front
Court, (4) Airy's chronograph (*Bottom*) (5) Gate clock, (6) Shepherd Mean

Solar standard clock (time-ball apparatus right). From *The Graphic*, 8
August 1885.

Fig. 127. THE CLOCK ROOM ABOUT 1930
Halley's old Quadrant Room. *(Left to right)* (1) Dent No. 2016,
controlled by Shortt No. 16 for B.B.C. time signals; (2) Dent No.
2, previously controlled by Cottingham – Riefler; (3) Shortt No.
11 (sidereal) slave clock.

g. 128. THE RUGBY ROOM ABOUT 1930
ortt No. 16 controlling rhythmic time signal and (through Dent
o. 2012 or 2016) BBC time signals, in basement of NE Dome.
eft to right) (1) slave clock; (2) diminished ½-seconds Rugby trans-
itter; (3) master pendulum.

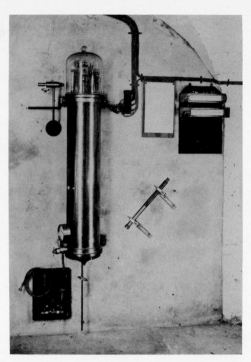

Fig. 129. SHORTT NO. 3 MASTER PENDULUM, SIDEREAL STANDARD, ABOUT 1930
Mounted in clock cellar in Flamsteed House basement.

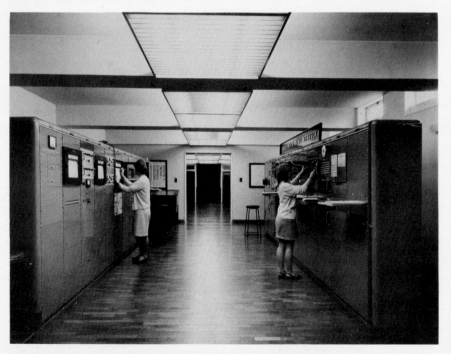

Fig. 130. TIME SERVICE CONTROL ROOM, HERSTMONCEUX IN 1967
Since July 1957, Britain's time-signals have originated from Herst-monceux. They are sent to the BBC by land line. The Control Room contains equipment for the reception and measurement of world-wide radio time-signals, and for the precise comparison of Herstmonceux's atomic clocks—with each other and, through the medium of the radio time-signals, with other atomic clocks at observatories all over the world.

Camera Obscura

Uffenbach reported that in 1710 there were two camerae obscurae which were "uncommonly pleasant on account of the charming prospect and the great traffic on the Thames".[21]

Halley's Minor instruments

Halley's telescopes, first recorded in 1721, were of 9, 15 and 24-foot focus, with wooden tubes and two micrometers of different forms.[22] The 24-foot refractor was of 2¼ inches aperture.[23] At the Visitation of 1726 a portable quadrant of the Royal Society was mentioned as well as the above telescopes.

The tubes of Halley's telescopes were removed from the Octagon Room in 1811 "to any place they can conveniently be put out of the way". At least one of Halley's micrometers was recorded in inventories up to 1930.

Bradley's Small Telescopes

When Bradley took over in 1742 he brought with him a 15-foot square wooden tube which could be used with 15-foot and 7-foot OGs. Later, a 10-foot OG was acquired (which seems never to have been used). He had a micrometer which is recorded as being used both with the 15-foot and 7-foot OGs.

Maskelyne recorded that the precise focal length of the 15-foot glass was 15 feet 3 inches, with an aperture, limited by paper, of 2·46 inches; eyepieces gave magnifications of 97, 64 and 48. The equivalent figures for the 7-foot glass were 7 feet 3·65 inches focal length, aperture 1·83 inches, magnifications 46, 30, 23. The 15-foot and 7-foot telescopes were used regularly in the Great Room from 1742 to 1760. The micrometer was still mentioned in the inventories of the 1930s.

In his re-equipment bill of 1748 to the Admiralty, Bradley asked for £20 for the 15-foot telescope and micrometer. He also acquired a 20-foot refractor by Bird for £7 10s. and a 6-foot Newtonian reflector by James Short for £100.

No observations are recorded with the former and it was ordered to be disposed of in 1774. The delivery of the latter was considerably delayed and Short lent a 6-foot Gregorian in 1751. After some dispute, the Newtonian was delivered in 1756. Once it arrived, it proved a great success (largely for Jupiter's satellites) until 1785 when it was superseded, nominally by Herschel's 7-foot

Newtonian, in practice by Dollond's 5-foot achromatic. Illustrated in Fig. 105, it had an aperture of 9·4 inches, a focal length of 6 feet 2·2 inches and four eyepieces magnifying between 274 and 74 times.

In 1820, Pond suggested it should be sent to the Cape but this probably did not happen. There was an unconfirmed report in 1830 that it was in private hands in Greenwich.

The instruments to be used by the various British expeditions to observe the Transit of Venus of 1761 were standardized by the Royal Society expedition-planning committee, one of the members of which was the Astronomer Royal, Bradley.

A 2-foot Gregorian by James Short, of 4·36 inches aperture[24] with an object-glass micrometer was lent to the Royal Observatory for the 1761 observations, and remained at Greenwich on loan until 1821, being used by Maskelyne in the 1769 transit. No other observations are recorded after 1765.

The 40-inch Movable Quadrant

Halley had asked for a movable quadrant in 1726 but funds ran out before it could be provided. Bradley's memorial to the King in 1748 asked for one by Bird, for £200. Delivered about 1754, this was virtually a scaled-down version of Graham's 8-foot mural quadrant mounted so that it could be turned to any azimuth.

Exactly what it was used for is not clear, nor is there any detailed description of it though both Jean Bernoulli, the Berlin astronomer, and Thomas Bugge from Copenhagen gave some details when they saw it in 1769 and 1774.[25] At first, it was kept in the Great Room but Bliss had a special observatory built on the site of Flamsteed's observatory in 1762. In 1797 it was lent to Trinity College, Dublin, for the new Royal Observatory at Dunsink. Its present whereabouts is not known.

Maskelyne's Refracting Telescopes

On 25 September 1765 there occurred at Greenwich an event not unimportant in the history of astronomy—an observation of the eclipse of one of Jupiter's satellites by "a 3½-foot telescope of Dollond with 3 object glasses".[26] This was the first recorded observation at Greenwich by an achromatic telescope (invented—though there is some dispute on this score—by John Dollond (1706–61) in 1758) taken with the first triple achromatic ever made by John's son Peter

(Fig. 106), to a design which was to prove so successful both technically and commercially over the next forty years.[27]

There is no record of payment by the Observatory for this telescope, so one presumes that Peter Dollond either lent or presented it to Maskelyne, so that he, Maskelyne, could give this prototype a thorough trial.

Dollond's investment paid off: the telescope proved so successful that hundreds were sold all over Europe. From that moment, all other refractors—even those on the large quadrants and transit instruments in observatories—became obsolescent. Probably generally similar to its successor illustrated in Fig. 32, it was used in the Great Room or in the open air, 39 observations being recorded between 1765 and 1772.

In 1772 a replacement was ordered from Peter Dollond for £63. This new 46-inch triple achromatic, by far the most successful of the RO's small telescopes, was used regularly from January 1774 until the Sheepshanks was mounted in 1838, 1084 observations of eclipses, occultations, etc., being recorded in those 64 years. Thereafter, three or four observations a year are recorded up to 1860. It is possible that it was used in Hawaii for the 1874 Transit of Venus.

This famous telescope, illustrated in Fig. 32, had an aperture of 3·6 inches, eyepieces magnifying 204, 132, 74 and 34 times, a large achromatic object-glass micrometer, and variable square apertures to measure the intensity of the light of Jupiter's satellites. It was kept in the Great Room from 1774 until 1780 when it was moved to the Advanced Room.

The wooden tripod stand, with a new polar axis and fork mounting but no telescope, is now preserved in the NMM. The whereabouts of the telescope is not known.

Maskelyne had several other refractors, only one of which was used to any extent—a 5-foot double-achromatic of 62 inches focus and 4¼-inch aperture (later reduced to 3·8 inches), probably seen top left in Fig. 105, apparently lent for trial by Dollond in 1785 when it superseded the 6-foot reflector by Short in the Advanced Room. In 1797 it was purchased for £70, remaining in use until 1849 when its object-glass was appropriated as a collimator for the transit circle.

The less successful—or less used—refractors were: (*a*) a 10-foot Dollond achromatic first referred to in 1769, probably lent for trial, illustrated in Fig. 107, last heard of in 1785; (*b*) Maskelyne's own

30-inch achromatic, fitted in 1780 with his newly invented prismatic micrometer,[28] on inventory until 1840; (*c*) a new 10-foot achromatic with a doublet of 5 inches diameter "which, Mr Dollond told me, is the largest and only one of that size he ever made",[29] proposed in 1793 and bought in 1797 for 150 guineas: its OG was appropriated by Pond for Troughton's 10-foot transit instrument in 1816 and subsequently used by Airy for the RZT in 1851—where it can still be seen.

The Reflectors of Maskelyne and Pond

Of the telescopes inherited from his predecessors, only Short's 6-foot Newtonian was of any real value and Maskelyne continued to use it regularly, at first in the Great Room, then, from 1780, in the Advanced Room. 184 observations were recorded between 1765 and 1785.

Maskelyne brought with him to Greenwich an 18-inch Gregorian by Short of $4\frac{1}{2}$ inches aperture. Some time before 1778, it was converted by Dollond into a 2-foot Cassegrain.[30]

The following is an extract from a letter of 3 June 1782 from William Herschel (who had discovered the Georgian planet, Uranus, the previous year) to his sister Caroline:

> "These two last nights I have been star gazing at Greenwich with Dr. Maskelyne and Mr. Aubert. We have compared our telescopes together, and mine was found very much superior to any of the Royal Observatory. Double stars they could not see with their instruments I had the pleasure to shew them very plainly, and my mechanism is so much approved of that Dr. Maskelyne has already ordered a model to be taken from mine and a stand to be made by it to his Reflector. He is however now so much out of love with his Instrument that he begins to doubt whether it deserves a new stand."[31]

At the Visitation the following year Maskelyne recommended purchasing a 7-foot Newtonian from Herschel. This was promptly supplied at a cost of £105 9*s.* 6*d.*, all concerned praising the new telescope at the expense of the old ones. Be that as it may, it was 1788 before the first record of an observation appeared, and only nine observations are recorded between 1799 and 1832, as against 541 for the 46-inch and 87 for the 5-foot refractor. That Herschel himself was a Visitor from 1784 to 1809 may be relevant.

In 1874, Airy offered to send the stand to Professor Pritchard in Oxford.[32] In 1930 the tube and mirror were presented to the Adler Planetarium, Chicago, where it is now displayed on a replica stand.

In 1813 a 10-foot Newtonian with a 9-inch speculum was ordered from Herschel at a cost of £350. Kept in the Great Room until about 1832, a single observation was recorded in 1829. The stand was probably disposed of in 1874: the tube and objective were in the inventory of 1933, and the latter is now displayed in Herstmonceux Castle.

A 25-foot Herschelian telescope of 15 inches aperture by John Ramage of Aberdeen (1783–1835), fully described and illustrated by Pearson, was mounted in the front courtyard in 1826. A particular feature was that the telescope could be managed by one person. Never a success, it was dismantled in 1836 after Ramage's death and returned to Ramage's executors.

All that are known to have survived of the small telescopes so far described are (a) the stand of the 46-inch refractor at the NMM; (b) the tube and objective of the Herschel 7-foot reflector at the Adler Planetarium, Chicago; and (c) the objective of the Herschel 10-foot reflector at Herstmonceux.

Maskelyne's Camera Obscura

First mentioned in 1788 in Fig. 7, it is not known when this camera obscura was installed in the western turret. It was removed in 1840 to make way for the Osler anemometer.

Pond's Fixed Telescopes

Dr Brinkley, later Bishop of Cloyne, reported that, according to the Dublin mural circle, the stars α Aquilae and α Cygni showed parallaxes of the order of 2″ to 3″. As this was not confirmed by the Greenwich circle observations, Pond had Dollond make two 10-foot achromatic telescopes of 4 inches aperture, one of which he fixed on the back of the circle pier directed to α Aquilae, one on the quadrant pier directed to α Cygni (Fig. 108).

Observations at Greenwich from 1816 to 1825 measuring the distances between the pairs *l* Pegasi and α Aquilae and between β Aurigae and α Cygni proved that no parallaxes of such magnitude exist.

The telescopes are now on display in the ORO.

Some Famous Instruments

Three instruments famous in their own right made somewhat fleeting appearances at Greenwich during the nineteenth century.

117

They are here described in the order in which they appeared at Greenwich.

The first was the 'Lee Equatorial', the famous 6-inch refractor by Tulley which Admiral W. H. Smyth used to compile his *Cycle of Celestial Objects*.[33] It can be seen in the background of Fig. 79.

With a 6-inch object-glass of $8\frac{1}{2}$-foot focus (described by Smyth as "the finest specimen of the late Mr Tulley's skill"),[34] Smyth had bought this telescope from Sir James South in 1829 and installed it in his observatory at Bedford, being mounted by George Dollond on a long mahogany polar axis in a similar manner to the Sisson sectors at Greenwich (p. 81). The driving-clock — said to be the first in England—was designed by Sheepshanks.

About 1840, Smyth presented this telescope to his colleague Dr John Lee (1785–1866) who erected it at Hartwell in Buckinghamshire. In 1869 it was bought by the Government from Lee's executors to observe the forthcoming Transit of Venus, being used successfully in Egypt in 1874.[35] From 1888 to 1914 it was at Hong Kong Observatory. In 1929 it was presented to the Science Museum, London, where it is now on display (Inv. No. 1929–949).

The second was the $3\frac{1}{2}$-inch Kew Photoheliograph, made for the Royal Society in 1854 by A. Ross to the design of Warren De La Rue. It was lent to the Royal Observatory in February 1873 and erected in a hut on the South Ground. From April 1874 to September 1875 the Kew instrument took daily photographs of the Sun at Greenwich, while the new Dallmeyer 4-inch photoheliographs— for which the Kew instrument had served as a prototype—were abroad on the various transit expeditions. It was returned to Kew in 1876 and presented to the Science Museum, London, in 1927 (Inv. No. 1927–124).

The third instrument was the Lassell 2-foot reflector. Completed in 1847 for William Lassell of Starfield near Liverpool, this was the Newtonian reflector of 24-inch aperture and 20-foot focal length[36] with which Lassell discovered Neptune's satellite Triton in 1846 and Saturn's eighth satellite Hyperion in 1848. He later used it with success in Malta. It was presented to the RO in 1883 by the Misses Lassell and mounted on its fork mounting (a very early example) on the South Ground in a new building surmounted by a new 30-foot dome by Cooke & Sons of York. The first recorded observation at Greenwich took place on 4 October 1884.

The telescope was not a success in its new site, however, and it was dismounted on 20 April 1892, being replaced on its mounting by the 12·8-inch refractor just dismounted from the south-east equatorial dome.

The dome was eventually used for the Thompson equatorial, erected on the central tower of the New Observatory from 1897. Lassell's mirror was presented to the Liverpool City Museum in 1956.

Portable Instruments and
the Transit of Venus, 1874 and 1882

In 1866 the Observatory acquired two portable instruments for use on expeditions—a transit instrument by Brauer of St Petersburg and an altazimuth with 15-inch circles by Troughton & Simms (called the 'brass beast' by the 1874 New Zealand expedition), both with diagonal reflectors and eye-view in the axis. The first has been at the Science Museum, London, since 1914, the second at the NMM (Fig. 109) since 1973.

In 1869 it was decided to fit out five British expeditions to observe the Transit of Venus of 8 December 1874, the instruments to be used again for the transit of 6 December 1882. The 1874 expeditions were organized by the Astronomer Royal.

A few of the instruments were borrowed but most were purchased and became the property of the RO:

(a) *portable transit instruments:* five new instruments, 'A' to 'E', each signed: *Troughton & Simms London 1870*, with telescopes of 36½-inch focus and 3-inch aperture, described in pp.40–41: 'B', 'C' and 'D' used for time-determination at Greenwich 1926–57, 'C' now lent to NMM, 'A' at Cape Observatory, 'E' sold to Egypt, 1929; (see Fig. 111)

(b) *portable altazimuths:* the axis-view instrument already described plus four new instruments 'A', 'B', 'C' and 'E', each signed *Troughton and Simms London 1870:* two 14-inch silver vertical circles divided to 5': transit telescope 20-inch focus, 2-inch aperture: 12-inch azimuth setting circle:[38] 'A' at Imperial College, 1929, 'B' at Cape, 'C' lost at sea 1882, 'E' to NMM 1973 (Fig. 110);

(c) *clocks:* twelve new clocks by Dent, described on p. 139–40;

(d) *photoheliographs:* five 4-inch × 5-foot by Dallmeyer, described on pp. 92–4;

(e) *6-inch equatorials:* two new telescopes by Simms, two by Cooke, four second-hand telescopes named after original owners: disposition 1929 or later: Simms No. 1, RO; Simms No. 2, Imperial College London; Cooke No. 1, not known; Cooke No. 2, R.O.; 'Naylor', lost at sea 1882; 'Hodgson', guiding telescope to Thompson 30-inch reflector (p. 101); 'Corbett', finder telescope for 28-inch refractor (p. 97); 'Lee', Science Museum (p. 118).

Some 20th-century Instruments

The following, though not really 'minor' instruments, have not been described elsewhere because of space limitations.

(a) *Spectrohelioscope:* 1929, lent by Hale observatories: Mount Wilson type with Anderson prisms and Rowland grating ruled 568 grooves/mm at a scale of 1 mm=4A. Coelostat flats 14 and 11·4 cm diameter—for visual observations of Sun in monochromatic light (Hα).

(b) *Coelostat with Lyot filter:* 1953, by Cox, Hargreaves & Thomson OG 26·7 × 775 cm. Used in conjunction with Dallmeyer OG: for studies of Sun in monochromatic light (Hα).

(c) *Newbegin 6¼-inch refractor,* c. 1888 by Cooke, gift of A. M. Newbegin, 1947, with 22-foot dome: mounted at Herstmonceux 1949 with Dallmeyer photoheliograph attached.

(d) *Bamberg small broken transit:* borrowed from Royal Observatory, Edinburgh, in 1942: used for time-determination at Abinger from 1947 to 1956.

(e) *Danjon astrolabe:* 1956, horizontal rotary mounting for visual observations over almucantar at 30° ZD: for time and latitude variation determination.

(f) *Steavenson reflector:* given to Cape Observatory by Dr W. H. Steavenson, 1956, returned to RGO, 1973. Primary mirror 76 × 306 cm, Cassegrain focus f:15: to be used for photoelectric photometry.

NOTES

1. RGO MS. 15/176. Table to convert micrometer readings into arc-minutes at different focal lengths.

1 a. *Phil. Trans.*, 1667.

2. RGOMS., Baily, p. 341. Crosthwait to Sharp, 10 December 1720.

3. See exchange of letters between Flamsteed and W. Molyneux between 24 July 1703 and 5 December 1704 in RGO MS. 34 and SCRO D./M.1/1. There is a description in Southampton City Record Office D./M.1/1 f. 26^r.

4. Southampton City Record Office D./M. 1/1 80^r, Flamsteed to W. Molyneux, 8 July 1684.

5. RS MS. 4/55 translated from Latin.

6. RS MS. 243 (Fl), Lr. 21, Flamsteed to Towneley, 24 March 1677.

7. J. R. Tanner, Ed., *Descriptive Catalogue of naval manuscripts . . .* IV (1923), 280.

8. Hooke diary 14 June 1677 (Robinson, p. 229).

9. RGO MS. 1/91v.

10. *HC* I, 353.

11. RS MS. 243 (Fl), Lr. 28, Flamsteed to Towneley, 3 November 1677. See also RGO MS. 36/45 Flamsteed to Moore, 2 November 1677.

12. RGO MS. 4/55, translation from Latin.

13. R. Hooke, *Animadversions on the first part of the Machina Coelestis . . .* (1674), 45–78, and Tables 1 and 2.

14. RS MS. 243 (Fl), Lr. 35, Flamsteed to Towneley, 4 July 1678.

15. Southampton City Record Office D.M. 1/1 f. 72^r, Flamsteed to W. Molyneux, 29 March 1684.

16. Baily, 45.

17. *Ibid.*

18. RS MS. 243 (Fl), Lr. 42 and 43, Flamsteed to Towneley, 25 October 1679 and 22 November.

19. Southampton City Record Office D.M.1/1, f. 24^v. Flamsteed to W. Molyneux, 29 May 1682.

20. Baily, p. 346.

21. Quarrell and Mare, *London in 1710* (1934), pp. 22–3.

22. Table of revolves, Halley, *Astron. Obs.*, p. 4 of RAS copy.

23. RGO MS. 251 36^r.

24. *Ibid.*, f. 91^v.

25. Bernoulli, pp. 85–6: Royal Library Copenhagen, Ny Kgl. Saml. 377e 4°, f. 87.

26. N. Maskelyne, *Astron. Obs.*, I (1776).

27. RGO MS. 251/36^r: Bernoulli, pp. 87–8.

28. *Phil. Trans.*, LXVII (1777), 799; Pearson, pp. 197–9.

29. W. Kitchiner, *A Companion to the Telescope* (1811), p. 27.

30. RGO MS. 251/41^v.

31. Letter in possession of Mrs Shorland of Bracknell: microfilm at NMM.

32. RGO MS. 660, Airy to Alexander Herschel and Pritchard.

33. W. H. Smyth, *A Cycle of Celestial Objects*, 2 vols., 1844.

34. W. H. Smyth, "Description of an Observatory", *Mem. R. astron. Soc.*, IV (1831), 557.

35. G. B. Airy, *Transit of Venus 1874*, p. 290.
36. Described in *Mem. R. astron. Soc.*, 18 (1850), 15–18.
37. Airy, *op. cit.*, p. 489.
38. *Ibid.*, pp. 18–19.

10

Magnetic and Meteorological Instruments

THE scientific disciplines of magnetism and meteorology have always interested the astronomers at Greenwich, though in the early days the incentive to take observations—and not many were taken—was for the furtherance of natural philosophy generally rather than for the purpose of the Observatory itself.

Bradley, however, realized that astronomical results could be affected by the changing meteorological conditions, primarily atmospheric pressure, temperature and humidity. In 1750 he started the regular barometer and thermometer readings, a series which remained unbroken until 1956—the longest series anywhere in the United Kingdom.

The voyages of exploration at the end of the eighteenth and early nineteenth centuries focused scientific attention on geophysics, particularly geomagnetism which, through the magnetic compass, directly affected the art of navigation—very much the concern of the Royal Observatory.

In 1816, the Admiralty asked that regular observations of magnetic variation (declination), should be taken at Greenwich, the results being sent to the newly founded Hydrographic Office.* Soon after this, a Magnet House was built in the garden and Pond started observations in 1818. As we saw on p. 9, the Magnet House had collapsed by 1824.

One of the first actions of George Airy on becoming Astronomer Royal in 1835 was to set up a Magnetical and Meteorological Department with James Glaisher as Superintendent, and to arrange for the building of a new Magnet House—cruciform in shape and entirely of non-magnetic materials—land in the park on the South Ground being enclosed for that purpose. At the same time he fitted anemometers, etc., on the roof of Flamsteed House and a special

*RSMS 371/15, Barrow to the Royal Society, 13 March 1816.

hut for the dip circle to the south of the Magnet House.

In 1847, photographic registration was introduced to produce continuous records of three magnetic elements and the barometer—one of the earliest examples of the scientific use of photography (Fig. 113).

In 1863-4, the activities of the department were extended by the construction of a range of buildings to the south of the Magnet House and by the excavation of a basement where magnetic apparatus, more or less duplicating that on the ground floor, was set up in a room whose temperature variation was less than 6°F throughout the year.

In 1880, Airy's electrometer mast of 1840 was removed.

Christie's new observatory buildings of the 1890s contained iron and steel that affected the old Magnet House. A new enclosure was therefore made in the park about 300 yards east of the old buildings and a new magnetic pavilion (Fig. 114) built for the absolute measurements of the magnetic elements. The instruments for measuring the variations of the magnetic elements—which were not appreciably affected by iron and steel in the new buildings—remained in the old Magnet House. The Magnetic Pavilion was completed in September 1898. The standard thermometers and the rain gauges were also removed to the Magnetic Enclosure.

In 1914 it was decided that the magnetic element variation instruments, substantially the same as in 1840, should be replaced by modern instruments and the opportunity taken to install them in the Magnetic Enclosure in a new Magnetometer House.

In 1917 the last instrument, the electrometer, was moved from the old Magnet House which was demolished after 80 years of use.

Soon after World War I, Southern Railway plans for electrifying their suburban system forced the RO to find a new site for magnetic observations, free of artificial disturbance. The new site was near Abinger in Surrey where work on the new Magnetic Observatory buildings started in January 1924. Observations began at Abinger in February 1925 and ceased at Greenwich in May 1926.

After World War II, railway electrification in Surrey forced another change of site, this time to Hartland Point in Devon where building started in 1955, and was completed in October 1956. Abinger closed down in April 1957.

The instruments used at Greenwich and Abinger for magnetic and meteorological observations are listed in Appendix II.

11

The Astronomical Clocks
at Greenwich

11.1. *Flamsteed's Astronomical Clocks*

ANY solution to the longitude problem by astronomical means must be based on the assumption that, for practical purposes, the Earth rotates on its axis at a constant speed. Proving this assumption was thus an essential ingredient in the longitude problem. By good fortune, at the time the Royal Observatory was founded, the means of so doing had just come to hand—the pendulum clock, invented by Huygens as recently as 1657. Moore and Flamsteed determined that one of the first tasks for the new observatory should be to conduct experiments to obtain this proof, using clocks more accurate than had ever before been made.

Late in 1675, Moore ordered from Thomas Tompion (1638–1713) two very special clocks, the most radical feature of which was to be that the pendulums would be 13 feet long (beating every 2^s), hung *above* the movements. Wren designed the Great Room to accommodate these unusual clocks, the dials of which can be seen in Fig. 2 to the left of the door, the pendulums being suspended behind the panelling.

Both Robert Hooke and Richard Towneley[1] played a part in the conception of this unusual arrangement which seems to have been adopted to reduce the swing of the pendulum to a minimum, thereby almost eliminating circular error. The tail of the pendulum engaged in the upward-facing crutch (see Fig. 122) moves less than $\frac{3}{4}$ inches each 2^s. Other design features included a dead-beat pinwheel escapement invented by Towneley and a year-movement. What

125

was missing was any form of temperature compensation and any sort of case to keep out the dust.

Though there were many teething troubles, the required proof—Flamsteed's Equation of Natural Days—had been obtained by 1678, with the help of a telescope fixed on the balcony so that the transit of Sirius could be observed day and night throughout the year. The clocks were regulated to mean solar time.

Removed from the observatory by Mrs Flamsteed in 1720 (they had been given to Flamsteed by Jonas Moore) these two year-clocks still survive, one at the British Museum, the other at Holkham Hall in Norfolk, the property of the Earl of Leicester. Their full story and that of Flamsteed's other clocks are told elsewhere.[2]

The other clocks used by Flamsteed at Greenwich were as follows:

(a) the sextant clock with a 1ˢ pendulum of 1675, Flamsteed's own property, made by Tompion, seen in the shadowed part in Fig. 68: this clock has not survived;

(b) the third clock in the Great Room of 1676, seen to the right of the door in Fig. 2; though its very existence is sometimes doubted, there is some evidence that a movement made in Lancashire under Towneley's supervision with a 6-foot pendulum, the dial and hands being made by Tompion to conform to the design of Jonas Moore's Great Clocks, could have been this one;

(c) a portable clock of 1680, *horologium ambulatorium*, sporadic references to which are made between 1680 and 1690;

(d) a new Arc House clock of 1690, Flamsteed's own property, apparently to replace (a) above;

(e) the Degree Clock of 1691, made by Tompion for Flamsteed, with a $\frac{2}{3}$ˢ pendulum to read Sidereal Time in arc—the dial showed, say, 232° 46′ 20″ instead of the equivalent 15ʰ 31ᵐ 05ˢ—which has survived and is now in the restored Sextant House in the Old Royal Observatory.

11.2. *Halley's Astronomical Clocks*

One of Halley's first actions after being granted £500 to re-equip the Observatory (see p. 6 above) was to acquire a transit instrument which he set up on the north-west side of Flamsteed House "with a plain week clock to stand by it"—the clock being provided by George Graham at the cost of £5.

Before 1726 he acquired two more clocks from Graham, with dead-beat escapements and simple pendulums, for £12 each. One of these he placed in the Great Room, the other in the Quadrant Room.

As outlined below, the two last clocks remained in active astronomical use until the twentieth century and, after 250 years of continuous running, are today still keeping time to within a few seconds a week—surely a great tribute to a superb craftsman.

The following is a summary of the subsequent history. The clocks were designated 'Graham 1' and 'Graham 2' about 1840. All the clocks described subsequently have a one-second beat unless stated otherwise.

Graham Week Clock

1721	Bought from Graham for £5. Installed as Transit Clock.
1742	Refixed facing north-west by Bradley.[3]
c. 1748	'A week clock with a simple pendulum' moved to Great Room.
12 June 1769	Rev. W. Hirst of Inner Temple observed Transit of Venus with 'Dr Halley's little clock belonging to the Royal Observatory'.[4] (Identification not certain.)
c. 1774	Moved to Middle Room. (Assistant's Calculating Room.)
1786	Dr Halley's 'Short Transit Clock' cleaned by W. Combe.
1796	Described as a clock 'made by the celebrated Graham, which once served immortal Halley as a transit clock. The face, which resembles one described by Ferguson, is the only curious part of it'.[5]
1818	In Assistants' Calculating Room. Then disappears from record.

Graham 1 (month)

Probably the surviving clock in 18th century floor-standing case in the office of the Head of the Time Dept. (1975), signed *Geo. Graham London* on plate on dial and numbered *621* on the back plate.

1725	Bought for £12. To Great Room. Probably simple pendulum.

1743	Gridiron pendulum by Graham—£15 13s. od.
1766	Cleaned by Shelton (back of hour disk is inscribed: *J. Shelton—Apr 21st 1766*). To movable quadrant observatory. (Advanced Room south of Quadrant Room.)
1772	Ruby pallets by John Arnold.
1791	Re-cased, probably by Earnshaw.
1846	Dent fits mercurial pendulum (ex. clock Grimalde & Johnson). To Altazimuth Dome.
1899	To New Altazimuth Pavilion.
1926	To Lower Record Room.
1937	To Visitor's Room, Greenwich.
1906	In office of Head of Time Department, Herstmonceux, with mercurial pendulum.

Graham 2 (*month*)

Probably the surviving clock in modern case in Chronometer Workshop (1975), signed: *Geo. Graham London* on dial and numbered *675* on back plate.

1725	Bought for £12. To Quadrant Room. Probably simple pendulum.
1744	Gridiron pendulum by Graham, £10.
1766	To Great Room.
1772	Ruby pallets by Arnold.
1791	Re-cased, probably by Earnshaw.
1834	To Circle Room.
1849	To Quadrant Room passage.
1854	To Battery Basement under N.E. Dome alongside Shepherd Motor Clock.
1856	Occasional Observatory.
1872	Zinc and steel pendulum by Dent.
1874	At Rodriguez I., Transit of Venus (T of V).
1891	To new Transit Pavilion. Sent on longitude expeditions to Paris in 1892 and Malta in 1909.
By 1909	In Thompson Dome.
1919	To Rosyth for Time-Ball.
1926	To Greenwich, Lower Record Room.
1960	In Chronometer Workshop, Herstmonceux, with modern case and zinc and steel pendulum.

11.3. *Bradley's Astronomical Clocks*

In 1748, Bradley asked for "a clock to be placed near the Transit Instrument of the same sort as that now standing by the Mural Quadrant".[6] This was provided by Graham, being made under his supervision by John Shelton. This clock remained as Transit Clock until 1821; was used for dropping the Time-Ball from 1833 until about 1924, and has been gracing the private room of successive Astronomers Royal since 1924. At the same time, Graham provided a 'chamber alarum' for £3. This is probably No. 667—a lantern timepiece alarm with verge escapement, in private hands since 1932.

Graham 3 (*month*) (Fig. 123)

Signed on dial *Geo. Graham London*. Movement not numbered. Floor-standing case. Present mercurial pendulum signed: *Richardson Royal Observatory*.

1750	Bought from Graham for £39. Steel pallets. Gridiron pendulum. To Transit Room, fixed to south wall, east of meridian openings.
July 1771	Ruby pallets instead of steel, by John Arnold.
February 1779	Steel escape-wheel instead of brass; 'perpetual ratchet' maintaining power instead of former machine; jewelled pivot holes to escape-wheel and pallets; pendulum support strengthened—all by Arnold.
January 1780	Moved to new stone pier close south-west of transit instrument (Fig. 32). Pendulum fixed to pier independent of clock case.
July 1789	Solid brass bob instead of brass-covered lead; supported at centre instead of bottom; special gridiron regulator fitted; small regulating ball fitted at bottom so that bob did not have to be moved for fine adjustment; 24^h hour-circle instead of 12^h—all by Larcum Kendall.
August 1792	Regulating ball removed.
June 1793	Motion work for hour hand reduced from three wheels to one; three cross-stays to pendulum instead of single one—by Earnshaw.
1821	Removed to Chronometer Room. Superseded as Transit Clock, first by Molyneux & Cope, then Johnson, finally (1823) by Hardy.

1828   To Quadrant Room for pendulum experiments by Sabine and Kater.

1833   To passage outside Great Room for dropping time-ball. Adjusted to mean solar time.

1836   Mercurial pendulum instead of gridiron; made to go a week; other changes unspecified.

March 1856 To new Ball Lobby at foot of Octagon Room stair-case. Alongside Shepherd Motor Clock for facilitating its regulation.

1914   To New Clock Room in west part of old Quadrant Room.

c. 1926 To Astronomer Royal's room.

1948   To Herstmonceux. Mercurial pendulum.

11.4. *Maskelyne's Astronomical Clocks*

In 1766 two assistant or secondary clocks (later known as journeyman clocks) were provided by John Shelton for £12 the pair. With wood-rod seconds pendulums, reading minutes and seconds only, with a loud beat and striking at the end of each minute, they could be set to beat in coincidence with the principal clock, either to assist in comparisons between clocks or, in windy weather, to reinforce the beat of the Transit Clock which was very soft.

Two similar journeyman clocks were purchased from Shelton in 1773. None of these four have survived, disappearing from the inventory in Airy's time. A journeyman clock by Shelton from the Radcliffe Observatory, preserved in the Museum of the History of Science at Oxford, is probably identical.

In 1772, Maskelyne asked for two new astronomical clocks for use with Sisson's equatorial sectors in the new terrace observatories. Made by Arnold, they were in all respects similar to the existing Graham clocks except that they had ruby pallets from the beginning and were somewhat surprisingly fitted with cycloidal cheeks at the suspension, a facility removed in 1836. Both were sold in the 1930s and are now in private hands.

In 1791 the Board of Longitude bought from Thomas Earnshaw: "a capital month regulator in mahogany box with gridiron pendulum; nine bars; jewel pallats and jewelled small wheel-holes—£89 5s."[7] Used by Vancouver off N.W. America in 1791–5 and by Flinders in Australia in 1801–2, this clock was kept at Greenwich when not otherwise engaged. It seems to have been taken over by the Royal Observatory when the Board was abolished in 1828. It was sold in 1938 and is now in private hands.

In 1809 Maskelyne ordered from William Hardy a clock for use with the new mural circle, fitted with Hardy's special escapement. Maskelyne seems not to have obtained a proper quotation for this clock because, when it was delivered in 1811 after Maskelyne's death, the Board of Ordnance refused to pay the £325 asked. The final sum paid seems to have been 200 guineas.

Clock Hardy became the Transit Clock in 1823 and continued in active service with successive transits until 1954. It is still going in its alcove to the south of Airy's Transit Circle.

Arnold 1 (month)

Signed on dial: *No. 1/John Arnold London.* 9-bar gridiron pendulum (1967). Floor-standing case.

1773	Ordered with Arnold 2. Whole cost £84 16s. To Quadrant Room.
1781	To East Dome.
1836	Cycloidal cheeks and old suspension spring removed by Dent.
1864	Seconds contact springs fitted. Reserve transit clock.
1874	To Sheepshanks dome in place of Earnshaw sent to T of V.
1876	To Upper Chronometer Room.
1909	To Thompson Dome to replace Graham 2 sent to Malta.
By 1926	New Chronometer Room.
1932	Sold to Mr P. Webster.
1975	In private hands in England.

Arnold 2 (month)

Signed on dial: *John Arnold No 2 London.* Floor-standing case. 5-bar gridiron pendulum (1967).

1773	As Arnold 1. To Great Room.
1775	To West Dome.
1836	Suspension spring and cycloidal cheeks removed by Dent.
1840	To bottom of Zenith Sector Apartment, Flamsteed House.
1848	To Quadrant Room passage.
1872	Zinc and steel pendulum.

1879	At Burnham, New Zealand, T of V.
1894	To University Observatory, Oxford.
1936	Returned to Greenwich.
1938	Sold to Messrs Clowes & Jauncey.
1975	In private hands in U.S.A.

Earnshaw (month)

Signed on dial: *Earnshaw London*. Wall-mounted 'waisted' case. 9-bar gridiron pendulum (1970).

July 1791	Purchase by Board of Longitude for £89 5s. Portable stand also accommodating journeyman clock.
1791	On board H.M.S. *Daedelus* with William Gooch. Transferred to H.M.S. *Discovery*, Capt. George Vancouver, exploring N.W. America.
1795	Back in Greenwich.
1801	After repair by Earnshaw, to H.M.S. *Investigator*, Capt. Matthew Flinders, landed several times during exploration of Australia.
By 1805	Back at Greenwich.
1828	Board of Longitude abolished. Clock transferred to observatory.
1834	To bottom of Great Zenith tube (Flamsteed House).
1839	To Sheepshanks Dome.
1854	To Harton Colliery, S. Shields, for pendulum experiments.
1856	Pendulum experiments at Greenwich. Then to Sheepshanks Dome.
1872	Zinc and steel pendulum.
1874	At Kaiya, Hawaii, T of V.
1880	To Sheepshanks Dome.
1938	Sold to Messrs Clowes & Jauncey.
1975	In private hands in England.

Hardy (week)

Inscribed on dial: *Willm Hardy London: Inv^t. et Fecit. New Dead-Beat Escapement by Dent*. Original case had round hood. No hood since 1850.

| 1809 | Ordered by Nevil Maskelyne. |
| 1810 | Case erected in Circle Room. |

1811	Clock installed. The cost of £325 was questioned, the final payment made being 200 guineas.
15 June 1812	Brought into use with Troughton mural circle.
1819	Use of mural circle for observing transits discontinued.
1823	Moved to Transit Room and attached to clock pier. Used as transit clock from 5 November.
1830	After the rate had been unsteady for some time, the Hardy escapement was removed and a dead-beat escapement substituted by E. J. Dent, Hardy himself being indisposed at the time. Hardy complained to the Royal Society later because Dent had been allowed to put his name on the dial.
1836	Jewelled holes removed by Dent. Pivots now turn in brass holes.
1850	Movement taken out of wooden case and mounted in Transit Circle Room in pit as transit clock.
1854	Contacts fitted to escape-wheel arbor to give galvanic impulses to the new chronograph every second (except the 60th in each minute). 'Hardy' thus became Sidereal Standard Clock.
1871	Clock 'Dent, 1906' became Sidereal Standard Clock and provided impulses for chronograph. Clock 'Hardy' now became reserve Sidereal Standard; it remained in position and was used for 'eye and ear' transit observations where the chronograph was not used.
1954	Last regular observation with Airy transit circle. Clock 'Hardy' still in position as transit clock (1975).

11.5. *Pond's Astronomical Clocks and Time-Ball*

An unusual clock was presented with the Shuckburgh equatorial in 1811—a three-month regulator by Arnold & Son with a $\frac{2}{3}$s (10″) gridiron pendulum and a dial indicating sidereal time in degrees, minutes and seconds of arc instead of the usual hours, minutes and seconds of time,[8] almost identical with Flamsteed's degree clock by Tompion of 1691. Like Flamsteed's, it seems never to have been used for its intended purpose at Greenwich. It is preserved by the National Maritime Museum (Fig. 124).

About 1812, Pond's assistant T. Taylor invented a very special alarm clock, designed to alert the astronomer some 3 or 4 minutes before the transit of a particular star, the names of the stars being engraved on the dial so that a pin could be inserted as appropriate (Fig. 125). It also functioned as an ordinary alarm clock. At least two of these were purchased and still survive at the National Maritime Museum.

About 1818, experiments sponsored by the Royal Society to ascertain the length of the seconds pendulum were carried out at Greenwich by Capt. Kater. The clock used in conjunction with the Kater invariable pendulum, made by Grimalde & Johnson, was bought by the Royal Observatory in 1819.

Two journeyman clocks were purchased from Johnson about 1818 but are otherwise unrecorded. A clock from the RO described as a journeyman clock now in the National Maritime Museum—with a painted deal case, a painted dial, wood-rod pendulum, but otherwise a conventional dead-beat movement, now with seconds contacts—could be one of these.

About 1829, Dent provided a new clock with mercurial pendulum for use in the Chronometer Room, seen in Fig. 127.

In 1833, the Admiralty issued the following Notice to Mariners:

> The Lords Commissioners of the Admiralty hereby give notice, that a ball will henceforward be dropped, every day, from the top of a pole on the eastern turret of the Royal Observatory at Greenwich, at the moment of one o'clock P.M. solar time. By observing the first instant of its downward movement, all vessels in the adjacent reaches of the river, as well as in most of the docks, will thereby have an opportunity of regulating and rating their chronometers. The ball will be hoisted half-way up the pole at five minutes before one o'clock, as a preparatory signal, and close up at two minutes before one.[9]

One o'clock was chosen because, at noon, the astronomers might be busy observing the meridian transit of the Sun.

This time-ball, the world's first visual time signal, was the brainchild of Capt. Wauchope, R.N. A contemporary description of its operation was as follows:

> The ball which is five feet in diameter, is a frame of wood, covered with leather and painted: it is perforated for the mast to pass through as it slides up and down. The shape of the mast is shown in fig. 4, supposing it to be sawn quite across, the triangular piece a is

moveable in an angle formed in that which is actually the mast, namely, the part b. This moveable or sliding rod, is fixed at one end to the ball; and is the means employed to push the ball to the top of the mast, which is accomplished by a chain passing from the bottom of the sliding rod, over a pulley, c. (fig. 3) to a cylinder at d, where it is wound round by a person turning a winch. The moment the ball reaches the top of the mast, the lower end of the sliding rod, to which is attached a piston, (something like the plunger of a common squirt,) is caught and held up by two clips e; thus the ball is supported till the moment of one o'clock; a person then pressing on a spring at f releases the clips at e, and down comes the ball.

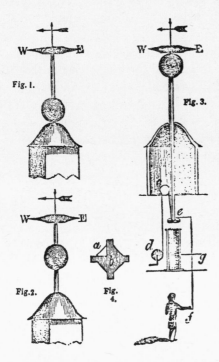

In order to prevent the ball dropping on the roof of the turret with violence, a cylinder g is placed under the piston, to receive it in descending; the air in the cylinder being thus compressed, and gradually forced by the descending weight out at an aperture near the bottom, allows the ball to come to rest very gently.

So satisfactory is the performance of the machinery, that the error of letting off the ball seldom amounts to three tenths of a second. The mast is surmounted by a wind vane, and a cross denoting the cardinal points of the compass.[10]

The apparatus remains substantially unchanged today except that, since 1852, the actual moment of drop has been controlled by an electric current from the Shepherd Mean Solar Standard clock and its successors, and, since 1960, the raising of the ball has also been made automatic.

The Ball clock used to time the dropping was first Arnold 2, then Graham 3, which remained in the Ball Lobby in Flamsteed House alongside clock Shepherd for many years as stand-by after the dropping was automated in 1852.

The ball was blown down into the Courtyard on 3 December 1855. The old wood-and-canvas ball was replaced by one of aluminium in 1919. Originally black, the ball was painted red some time in the 1920s.

Except on days when snow or high winds prevented raising, during the war years, and for some years in the 1950s, the ball has dropped at precisely 1 p.m. every day since 1833.

It is today controlled by a Synchronome clock in Flamsteed House basement: this clock, the responsibility of the Department of the Environment, is automatically corrected every 15 minutes by the 6-pip time signal sent from Herstmonceux to Greenwich by land-line via the B.B.C. Thus the time-ball today is controlled—albeit indirectly—by the world's most accurate atomic clocks.

Arnold & Son Degree Clock (3-month) (Fig. 124)
Inscribed on dial: *John Arnold & Son London*: $\frac{2}{3}$" gridiron pendulum. Floor-standing case.

c. 1793	Bought by Sir George Shuckburgh for use with Ramsden equatorial.
1811	Presented to RO.
1929	Transferred to Science Museum, London.
1967	Transferred to National Maritime Museum, Greenwich.

Grimalde & Johnson (8-day)

c. 1818	Used in Quadrant Room for pendulum experiments at Greenwich.
1819	Purchased by RO for £120.
1838	To new magnetic building as sidereal clock. Fitted with deal pendulum belonging to Baily. (Original mercurial pendulum was fitted to Graham 1 in 1846.)

1932	Sold to Percy Webster.
1967	Clock in private hands in Canada possibly this one.

Dent 2 (8-day) (Fig. 127)

1829	Clock by Dent with mercurial pendulum used by Sabine for pendulum experiments. Probably the same as that used hereafter in Chronometer Room for daily comparisons of chronometers.
1853	Chronometer comparisons with Shepherd Sympathetic dial in Chronometer Room. Dent 2 remains as reserve.
1869	New Chronometer Room.
By 1905	To basement of Magnetic Observatory. Contact springs fitted. Adjusted to sidereal time. For use when sidereal standard (Dent, 1906) is under repair.
1908	To Record Room. Remains reserve sidereal standard.
1924	Controlled by Cottingham-Riefler sidereal clock.
1938	Regulated to mean solar time. To Octagon Room.
1940	To Store.
By 1954	In Library at Herstmonceux Castle.

11.6. *Airy's Astronomical Clocks*

Airy's contributions to horology were many and varied. His improvements to chronometers do not concern us here; his innovations in clock Dent No. 1906 were important but, perhaps, not fundamental. It was for his time-distribution system of 1852—based upon the newly discovered 'galvanism', applied both to the clocks which gave the time signals and to the electric telegraph which distributed them—that we should best remember him.

Before the coming of the railways, each community kept its own local time. The railways, however, pointed the need for a standard time within the system—and the civilian population found this convenient too. Airy appreciated this need and, furthermore, realized that electricity provided the means of satisfying it.

The time-distribution system he set up for the United Kingdom in 1852 (immediately after the Great Exhibition), is in essence the same as that used world-wide today. The components have changed —time-keeping is by atomic instead of pendulum clocks, time-distribution by radio instead of by wire, the scope is world-wide instead of being limited to the U.K.—but the basic system remains the same as that devised by Airy in 1852.

His first clock acquisition was not, however, connected with this system. In 1846, a conventional regulator by Mudge & Dutton was presented by Rev. Charles Turnor.

The heart of Airy's time-distribution system of 1852 was a galvanic 'motor clock' made by C. Shepherd who had installed a not-very-successful system of sympathetic clocks at the Great Exhibition. Electricity was used not only as the driving force but also to provide seconds impulses to drive 'sympathetic' clocks around the observatory (including the famous gate clock) and at London Bridge station, hourly signals to the Electric Telegraph Company (later to the G.P.O.) and the South-Eastern Railway Co. whence time signals were distributed by telegraph as far afield as Guernsey and Dublin. From 1855 impulses from the Shepherd Clock at Greenwich dropped time-balls at Greenwich, Deal and London: time-guns were fixed automatically at several points in N.E. England.[11] A time-signal was sent to Harvard University by the Atlantic cable in 1866.

An important feature of Shepherd's clock was the facility for correcting the clock without touching the hands. At least once a day it was compared with the sidereal standard clock—whose error had been found astronomically—and if necessary re-set to the correct mean time.

Clock Shepherd became known as the Mean Solar Standard Clock; its opposite number in the system was the Sidereal Standard Clock—initially the transit clock Hardy which had, in 1854, had electric contacts fitted so that sidereal seconds impulses could be transmitted both to the Time Desk and to the new barrel chronograph by Dent fitted in the ground floor of the N.E. Dome in the same year[12] (Fig. 80 and 126), recording on a paper chart the impulses from the Sidereal Standard and the times of observation from the transit circle, the altazimuth and, later, the Great Equatorial.

In 1869, Airy ordered from Dent a new Sidereal Standard Clock, No. 1906, with a detached escapement analogous to the ordinary chronometer escapement and a very sophisticated temperature-compensated pendulum. In 1873, compensation for changes in barometric pressure was added. This clock, mounted in the basement of the Magnetic Observatory where the temperature remained almost uniform, performed excellently for many years, its accuracy being of the order of $0^s \cdot 1$ per day.

Among the new instruments ordered in 1870 for the Transit of

Venus expeditions of 1874 and 1882 (p. 119) were 12 identical regulator clocks from Dent—Nos. 1914, 1915, 1916, and 2009 to 2017 inclusive. As can be seen below, eight of them became observatory property.

A clock by Molyneux acquired for the 1874 transit also reverted to the RO.

Mudge & Dutton. Gridiron pendulum
    1846: presented by Rev. Charles Turnor through Admiral Smyth. 1851: to Reflex Zenith Tube room; sidereal.
    c. 1904: to Upper Chronometer Room, converted to Mean Solar. 1932: sold to Percy Webster; present whereabouts unknown.

Shepherd Motor Clock (Fig. 126)
Inscribed: *Shepherd Patentee 53 Leadenhall Street London*
    1852: installed in Galvanic Room on ground floor of N.E. dome. By 1855: as Mean Solar Standard clock, drives sympathetic clocks outside gate, in Chronometer Room, in Computing Room, in hall of Flamsteed House, in the Royal Hospital Schools and in London Bridge Station. Drops time-balls on Flamsteed House, in Deal and in the Strand, London. Sends hourly signals to telegraph and railway offices. 1856: moved to foot of Octagon Room Staircase; Graham 3 alongside. 1860: to Ball Lobby outside Octagon Room. 1866: time-signals sent to S.S. *Great Eastern* during the laying of the Atlantic cable, then to Harvard Observatory in October after its completion. 1893: superseded as Mean Solar Standard by Dent No. 2012.
    1914: with Graham 3, moved to new clock room in old Quadrant Room; at some stage later, moved to Upper Record Room. 1967: transferred to NMM; displayed in Frank Dyson Gallery in going order. 1974: to Spencer Jones gallery.

Dent No. 1906. Sidereal. Zinc and steel pendulum
Inscribed: *E. Dent & Co., 61 Strand, London, No. 1906, A.D. 1870*
    1869: Ordered from Dent. May 1871: fixed on north wall of Magnetic Basement. August 1871: brought into use as Sidereal Standard Clock. 1873: modified to provide compensation for changes in barometric pressure.
    September–December 1911: after overhaul, erected in new Clock Room in Old Quadrant Room; Dent 2 Sidereal Standard during

changeover. October 1922: superseded as Sidereal Standard by Cottingham-Riefler; remained in Clock Room. 1951: Festival of Britain exhibition, South Bank, London. By 1954: Chronometer Workshop, Herstmonceux. After 1962: mounted outside Time Service Control Room. 1974: lent to NMM; mounted in Spencer Jones gallery.

Dent No. 1914 (All the Transit-of-Venus clocks which follow are 8-day)
1874: at Cairo, Transit of Venus (T of V); then presented to the Khedive of Egypt.

Dent No. 1915
1874: at Kerguelen I., T of V. 1877: Sheepshanks dome. 1882: to Dent for overhaul; probably to S. Africa for Transit of Venus. 1884: sold to Natal Observatory.

Dent No. 1916. Zinc and steel pendulum
1874: at Honolulu, T of V. 1875: to Cape Observatory on loan. 1971: returned from Cape: mounted in Herstmonceux Castle, west entrance, first floor.

Dent No. 2009. Wood pendulum. Sidereal.
1874: at Cairo, T. of V. 1882: possibly to T of V. July 1894: to S.E. Equatorial dome.
By 1929: to Astrographic dome instead of Dent 2017. 1940: to Abinger to replace Dent 2012 as controlled clock for transmission of time signals. 1956: lent to Jeremiah Horrocks Observatory, Preston.

Dent No. 2010
1874: at Kerguelen I, T of V. 1882: at Jamaica, T of V, c. 1885: to Devonport for dropping of time-ball. 1974: still in Chart Depot, Devonport.

Dent No. 2011. Zinc and steel pendulum
1874: at Kerguelen I., T of V. 1882: possibly to T of V. 1886: lent to Kew Observatory.
1926: transferred to Kew permanently. 1970: transferred to Herstmonceux. 1973: acquired by NMM. 1974: displayed in Spencer Jones Gallery.

Dent No. 2012. Originally wood-rod, later zinc and steel pendulum.

1874: Honolulu, T of V. 1884: fitted with contacts for sending time signals as reserve Mean Solar Standard; mounted in Ball Lobby, given zinc and steel pendulum. May 1893: superseded Shepherd as Mean Solar Standard; driving sympathetic clocks and giving time signals.

November 1911: re-erected in New Clock Room. 1925: with Dent No. 2016, fitted with contacts for transmission of B.B.C. time signals. 1927: controlled as slave clock by Shortt No. 16. 1939: to Abinger for emergency time service transmitting time signals under control of Shortt MT clock. 1954: to Herstmonceux store. 1961: mounted in lobby of Equatorial Group building, regulated to sidereal time.

Dent No. 2013. Wood-rod pendulum

1874: at Honolulu, T of V. 1882: probably T of V. By 1895: lent to Cape Observatory.

Dent No. 2014. Wood-rod pendulum

1874: at Rodriguez I., T of V. 1893: Solar department; south wing of New Observatory.

By 1905: Lower Chronometer Room; Mean Solar. By 1929: Thompson dome. 1960: to Hartland Point magnetic station.

Dent No. 2015

1874: at Cairo, T of V. 1882: at Bermuda T of V. 1882: lost in R. Mersey in wreck of S.S. *City of Brussels*.

Dent No. 2016. Zinc and steel pendulum (Fig. 127)

1874: at Burnham, N.Z., T of V. 1882: at Barbados, T of V. 1891: contact springs for longitude operation; mean solar. 1893: Lassell dome. 1894: to Lower Chronometer Room. 1899: to Thompson dome to control driving clock of Thompson equatorial.

1922: contact springs as duplicate controlling clock for mean time circuits. 1924: controlling B.B.C. time signal. 1927: became slave clock to Shortt 16 for time signals. 1939: to Abinger for emergecy time service. 1940: replaced at Abinger by Dent 2009; sent to Edinburgh as slave to Shortt Nos. 11 and 16. 1948: to Greenwich Rugby Room under control of Shortt No. 67. 1956: to Chronometer Workshop, Herstmonceux. 1960: to lobby of Equatorial Group building, mean solar.

Dent No. 2017. Zinc and steel pendulum; Sidereal

1874: Burnham, NZ., T of V. 1882: Barbados T of V. 1891: to Astographic dome.

By 1929: in Cookson hut. 1960: to Hartland Point magnetic station.

Molyneux. Wood-rod pendulum

1874: at Rodriguez I., T of V; there is no record of where this clock came from. 1883: to Magnetic Observatory, northern arm.

By 1905: replaced Mudge and Dutton in RZT room. *c.* 1930: to S.E. Equatorial dome. 1957: set up in Herstmonceux Castle, Drummer's hall.

11.7. *Dyson's Astronomical Clocks*

During the time of Airy's successor, Christie, no astronomical clocks were acquired by the Observatory though the Greenwich Meridian was chosen as the basis for the world's time-zone system in 1884.

Early in the century, the leading observatories began acquiring clocks significantly more accurate for astronomical purposes than their predecessors, designed by Siegmund Riefler of Munich (1847–1912). Greenwich did not, however, acquire one until after World War I when, in 1921, Riefler No. 50 was purchased from the University of Manchester. This clock never seems to have been put to use at Greenwich and was sold in 1938 to Professor Schönland. In the same year, a clock with a Riefler escapement—with an airtight case, an invar pendulum, and a self-winding movement—was ordered from E. T. Cottingham, being brought into use in 1922, taking over as Sidereal Standard from Dent No. 1906 in October.

The Cottingham–Riefler clock, however, was to have a very brief spell as sidereal standard because of the advent in 1925 of the Shortt free-pendulum clock—a development in time-keeping as fundamental as the invention of the pendulum clock 250 years before.

The best previous clocks had an accuracy of about 1^s in ten days: the Shortts were accurate to 1^s in a year. Within a very few years, the free-pendulum clocks ousted for Time Service purposes all the older clocks, some of which had been in active astronomical use for nearly 200 years, none of which (except the two Rieflers), were less than 55 years old.

The free-pendulum clock was perfected in 1921–4 by W. H.

Shortt, working in conjunction with F. Hope-Jones and the Synchronome Co. Ltd. In ordinary pendulum clocks, the free swinging of the pendulum—on which timekeeping accuracy depends—is interfered with by the need to sustain the pendulum's motion and to count the swings to tell the time. In the free-pendulum clock, pioneered by R. J. Rudd in 1899, these two functions are carried out by a subsidiary or 'slave clock', the master pendulum swinging quite freely except for a fraction of a second each half-minute when it receives impulses from the slave.

Shortt No. 3 took over as sidereal standard in 1924, and a second sidereal clock, No. 11, was acquired in 1926. In 1927, a mean-time clock, No. 16, was purchased, being used to control the B.B.C. time-signal slave clocks, Dents Nos. 2012 and 2016, and to control the new Rugby rhythmic time-signals (started 19 December 1927) which, by using the vernier principle, enabled navigators at sea and surveyors on land to obtain time to an accuracy of $0^s\cdot01$. Special slave clocks, beating 61 and 122 to the minute ('diminished seconds' and 'diminished half-seconds') were used.

In passing, one must mention another development fundamental to the Time Service—radio time-signals. The earliest regular time-signals were broadcast from Washington in 1905. The B.B.C. six-pip time-signal, controlled from Greenwich, was first broadcast on 5 February 1924, though the chimes of Big Ben had ushered in the New Year of 1924 in the B.B.C.'s programme.

In 1925 the new Magnetic Station opened at Abinger, Surrey. The Standard clock, fitted with contact springs for the time-scaling of records, was one by John Shelton, probably the one originally purchased by the Board of Longitude for the 1769 Transit of Venus, for use by John Bradley (nephew of the former Astronomer Royal) at the Lizard.[13]

Shelton (month)
Gridiron pendulum. Inscribed on dial: *John Shelton London*.

1768: probably the clock purchased for £40 by the Board of Longitude for use by John Bradley observing the Transit of Venus at the Lizard in June 1769. 1770–1820: whereabouts not known; probably not one of Cook's clocks.

1820: possibly with Basil Hall in the *Conway* to the Pacific.

1902–14: Plymouth Chronometer Depot. 1922: transferred to RO by Hydrographer as spare clock in lieu of Graham 2 (in Rosyth).

1924: contacts and magnetic correction apparatus fitted by Kullberg. 1925–59: at Abinger as Standard Clock for scaling of records. 1963: mounted at Herstmonceux Castle near East Entrance outside Library.[14]

Shortt No. 3 (sidereal) (Fig. 129)

November 1924: master installed in clock cellar in Flamsteed House; slave in Clock Room (Old Quadrant Room). 1 January 1925: took over sidereal standard from Cottingham. 1926: second free-pendulum clock acquired (No. 11). 1929: overhauled and invar bob fitted. 1940: time service transferred to Abinger. 1940 or 1941: Shortt No. 3 to Royal Observatory, Edinburgh, for reserve time service. January 1946: overhauled and re-erected at Greenwich. About 1957: to Herstmonceux. 1967: to NMM; erected in Frank Dyson Gallery. 1974: moved to Spencer Jones Gallery.

Shortt No. 11 (sidereal)

July 1926: master and slave clocks mounted alongside those of No. 3. 1929–40: took over Sidereal Standard periodically from No. 3. January 1941: to Edinburgh for emergency time service. 1946: overhauled and re-erected at Greenwich. About 1957: to Herstmonceux. 1960: to Government Observatory, Sydney, N.S.W.

Shortt No. 16 (mean time) (Fig. 128)

1927: master and two slaves erected in Rugby Room (ground floor of Shuckburgh dome). December 1927: start of Rugby rhythmic time signal; Shortt No. 16 also controlled Dent Nos. 2012 and 2016 transmitting Post Office and B.B.C. time-signals. February 1935: dismounted for overhaul; new Shortt No. 49 took over. April: master moved to clock cellar under Flamsteed House adjacent to sidereal clocks. January 1941: to RO Edinburgh for emergency time service. January 1946: overhauled and re-erected at Greenwich. 1954: to Cavendish Laboratory, Cambridge. About 1957: to Herstmonceux. 1971: to NMM (Acq. No. 71–13) with 2 slaves. 1974: erected in Spencer Jones gallery.

11.8. *Spencer Jones's Astronomical Clocks*

In 1934 a second mean time free-pendulum clock, Shortt No. 49, was purchased, while, in 1930, Mr H. R. Fry presented his own No.

40—very refined with the slave clock jewelled throughout—as an additional sidereal clock.

By 1938, radio time-signals made it possible to compare clocks in different observatories all over the world, several times a day with great precision. This led to the abandonment by the U.K. Time Service of Airy's concept of specified standard clocks at Greenwich —sidereal and mean time—in favour of taking the mean of many clocks, initially five Shortts at Greenwich and one at Teddington later joined by one at Edinburgh and three at Abinger.

In the summer of 1939 some of the older clocks were sent to the Magnetic Observatory at Abinger, Surrey, to prepare for the setting up of a war-time emergency time-service, two more Shortt Clocks being ordered from Synchronome. By the autumn of 1940, when Greenwich was suffering from air raids, the new Shortt Nos. 66 (sidereal) and 67 (mean time), together with No. 61 (sidereal) lent by Mr Fry, were in full working order, and Abinger (known as Station A) took over the Time Service, the Greenwich clocks being taken out of service in December.

Early in 1941 a reserve Time Service was set up at the Royal Observatory, Edinburgh (known as Station B), Shortt Nos. 3, 11 and 16 being sent from Greenwich. Station B closed down in January 1946 when the three Shortt clocks returned to Greenwich. The headquarters of the Time Service remained at Abinger until 1957, never returning to Greenwich.

The development of the quartz crystal clock in the 1930s was no less fundamental than that of the free-pendulum clock in the 1920s. Greenwich acquired its first in 1939. By 1942, quartz clocks from Abinger, Edinburgh and Teddington had begun to form part of the mean clock. In 1943, the G.P.O.'s quartz clocks at Dollis Hall ousted Shortts as primary standards and, in 1944, the control of time signals was taken over by the new quartz clocks at Abinger, and geared phonic motors provided sidereal time. By 1946, 12 had been installed at Abinger, 6 at Greenwich.

Shortt No. 49 (mean time)

November 1934: installed in Flamsteed House clock cellar. February 1935: first took control of mean time clocks used for G.P.O. and B.B.C. November 1942: to new building Abinger. 1954: to clock cellar, Herstmonceux Castle. 1962: to Perth Observatory, Western Australia.

Shortt No. 40 (sidereal)

1932: property of H. R. Fry; regulated to mean time. 1936: presented to RO by H. R. Fry; regulated to sidereal and erected in Flamsteed House clock cellar. November 1942: to new building Abinger, with No. 66 (sidereal) and 49 (m.t.). 1943: to clock cellar, Herstmonceux Castle. About 1960: mounted outside time-service control room.

Shortt No. 61 (sidereal)

Autumn 1940: borrowed from H. R. Fry and set up in emergency time-station, Abinger. January 1946: returned to H. R. Fry on closing down of emergency time service.

Shortt No. 66 (sidereal)

Autumn 1940: erected at Abinger. November 1942: to new building Abinger with No. 40 (sidereal) and 49 (m.t.).

Shortt No. 67 (mean time)

Autumn 1940: erected at Abinger. January 1948: to Greenwich for control of Dent clocks for 6-pip time signal. 1958: to Cape Observatory.

11.9. *Woolley's Astronomical Clocks*

Just before Spencer Jones's retirement, there occurred yet another event of great importance in the history of timekeeping—the bringing into service of a caesium atomic clock at the National Physical Laboratory at Teddington. Almost at a stroke, quartz clocks became obsolete as primary standards. We said that the best clock prior to the 1920s was accurate to 1^s in ten days and the Shortt 1^s in a year. A maker of today's Caesium Beam Frequency Standard claims that it will gain or lose no more than 1^s in 3171 years. Variations in the rate of rotation of the Earth could be detected by quartz clocks; with caesium clocks, they can be measured with some accuracy.

The Time Service moved from Abinger to Herstmonceux in 1957, based on quartz clocks at Herstmonceux, Teddington and Dollis Hill, with comparisons with caesium standard at N.P.L. In June 1966 the first Caesium Standard was installed at Herstmonceux —a Model 5060A made by the Hewlett Packard Co., U.S.A. All quartz clocks were superseded by 1966. In 1974 six Hewlett Packard Standards formed the basis of the Time Service.

Conclusion

Though the spread of Greenwich Time is really outside the scope of this volume, we will nevertheless conclude this chapter—and the book—by making a brief review of achievements in the last 300 years—in time-determination, in timekeeping, and in time-distribution.

Then—in 1675—Flamsteed's equal-altitude observations for time-determination were probably accurate to about 5^s of time: now—in 1975—the PZT is accurate to ± 20 milliseconds for a single star, ± 4 milliseconds for a typical night's observations. Then, Greenwich time depended upon two clocks at Greenwich whose accuracy was probably of the order of 1^s in $3\frac{1}{2}$ hours: now, our time is based on a weighted mean of many clocks all over the world (upwards of 50 in 1974), each of which loses or gains no more than 1^s in about 3000 years. Then, Flamsteed proved the Earth rotated at a constant speed: now, we have proved that it does not.

Then, Greenwich Time was of interest only to the astronomers who worked there: now, it is the basis of the time kept all over the world.

NOTES

1. Richard Towneley (1629–1707), was head of a distinguished Roman Catholic family from Lancashire. He first propounded Boyles' Law, called by Boyle himself 'Mr Towneley's hypothesis'.
2. D. Howse, *The Tompion Clocks at Greenwich* ... (*Antiquarian Horology* reprint 1970–1).
3. Rigaud, pp. lii, lxx.
4. *Phil. Trans.*, 1769, 229.
5. J. Evans, *Juvenile Tourist* (1810), 333–5.
6. ROV Vol. I, 32.
7. RGO MS. 545/231, Board of Longitude, Vol. XIV.
8. G. Shuckburgh, *Phil. Trans.*, 1773, 87–9.
9. *Mechanics' Magazine*, XX, 536 (16 November 1833), 112.
10. Anon, "The Royal Observatory Greenwich", *The Weekly Visitor*, CXXI (17 February 1835), 67–8; see also *The Nautical Magazine*, October 1835, p. 584.
11. W. Ellis, "Description of the Greenwich time-signal System", *Astron. Obs.*, 1879, Appendix.
12. Anon, "Description of the Galvanic Chronographic Apparatus ...", *Astron. Obs.*, 1856, Appendix.
13. D. Howse, *The clocks and watches of Captain James Cook* (*Antiquarian Horology* reprint 1969).
14. *Ibid.*, pp. 285, 295–7.

Bibliography

A—*General descriptions of the observatory, arranged in chronological order*

1687   BENTHEM, Henrich Ludolff, 1694, *Engländischer kirch und Schulen staat* (Lüneburg), pp. 53–4.

1710   UFFENBACH, Zacharias Conrad V., 1753–4. *Mertwürdige Reisen Durch Niedersachsen, Holland und Engellend* II (Frankfurt & Leipzig) pp. 445–5 translated by QUARREL, W. H. and MARE, Margaret, 1934, in *London in 1710* (London: Faber & Faber), pp. 20–6.

1714   Royal Society Council meeting minutes 18 February 1714 (RGO MS 617/10–11).

1715 (say)   FLAMSTEED, John, 1725. *Historia Coelestis Britannicae*, III (London: 1725), Prolegomena, pp. 101–13.

1726   Royal Society Council minutes 2 March 1727 (RGO MS 617/18–24).

1726   WEIDLER, Io. Frider, 1727. *Io. Frider Weidleri de Praesenti Specularum Astronomicarum Statu dissertatio* (Wittenberg), pp. 3–11.

1769   BERNOULLI, Jean, 1771. *Lettres Astronomiques* (Berlin), pp. 77–100.

1776   MASKELYNE, Nevil, 1776. *Astronomical Observations made at the Royal Observatory at Greenwich . . . 1765 to 1774* (London), Preface i–xi.

1777   BUGGE, Thomas, 1777. MS diary (Royal Library, Copenhagen, Ny Kgl. Saml. 377e). Translation edited by K. Pedersen in preparation. Xerox of MS and translation at NMM.

1796–8   EVANS, John, 1810. *Juvenile Tourist*, pp. 332–43. Evans was Assistant at the RO 1796–8.

1829   REES, Abraham, 1829. *Cyclopaedia*, Vol. 25, article 'Observatory'.

1836–1908   AIRY, G. B. and CHRISTIE, W. H. M. Annual Volumes 1836–1908. *Greenwich Observations*. A detailed description of the instruments, etc., was contained each year in the Introduction.

1839   [AIRY, G. B.], 1839. *Penny Cyclopaedia*, pp. 440–2.

1845   [AIRY, G. B.], 1845. 'Plan of the Buildings and Grounds of the Royal Observatory, Greenwich with Explanation and History', *Greenwich Observations 1845*. Appendix.

1862   GLAISHER, James, and DUNKIN, Edwin, 1862. 'The Royal Observatory, Greenwich', *Leisure Hour*, 1862, pp. 7–11, 22–6, 39–43, 55–60.

1863   [AIRY, G. B.], 1862. Plan of the Buildings and Grounds of the Royal Observatory, Greenwich, 1863, August; with Explanation and History', *Greenwich Observations 1862*. Appendix II.

1863   ANON, 1863. 'A Night at Greenwich Observatory', *Cornhill Magazine*, March 1863, pp. 381–9.

1866   CARPENTER, J., 1866. 'John Flamsteed and the Greenwich Observatory', *The Gentleman's Magazine*, February, March, April 1866, pp. 239–52, 378–86, 549–59.

1872–8   WALFORD, Edward, *Old and New London*, Vol. 6, pp. 212–23.

c. 1879   DUNKIN, Edwin, c. 1879. *The Midnight Sky* (London: The Religious Tract Society).

1887   WINTERHALTER, A. G., Lt. USN, 1889. 'The International Astro-photographic Congress and visits to certain European observatories ...', *Washington Observations 1885*, Appendix I (Washington DC, Government Printing Office), pp. 143–63.

1898   MAUNDER, E. W., January–October 1898, 'Greenwich Observatory', *Leisure Hour*, pp. 152–60, 228–38, 293–7, 376–9, 561–5, 642–6, 695–702. A slightly fuller version of the next entry.

1900   MAUNDER, E. Walter, 1900. *The Royal Observatory Greenwich. A Glance at its History and Work* (London: The Religious Tract Society).

1943   SPENCER JONES, Sir Harold, 1943. *The Royal Observatory Greenwich* (London: The British Council).

B—Primary sources not mentioned in A

HC—FLAMSTEED, J. 1725, *op. cit.*

HOOKE diaries—ROBINSON, H. W. (Ed.), 1935. *The Diary of Robert Hooke 1672–1680* ... (London: Taylor & Francis).

KITCHINER, W., 1811. *A Companion to the Telescope.*

LALANDE, Jerome Français de, 1771. *Astronomie* (Paris), 2nd ed. II.

LOCKYER, J. Norman, 1878. *Stargazing Past and Present* (London: Macmillan & Co.).

Astron. Obs.—Greenwich Observations—annually since 1742 by Bradley, Bliss, Maskelyne, Pond, Airy, Christie, Dyson, Spencer Jones and Woolley.

PEARSON, William, 1829. *An Introduction to Practical Astronomy*, II (London).

Phil. Trans.—Royal Society of London. *Philosophical Transactions* annually.

PRO—Public Record Office, London.

Report—Report of the Astronomers Royal, annually from 1836 to 1964, printed with *Greenwich Observations.*

RGOMS—Royal Greenwich Observatory manuscripts.

ROV—Reports of the Board of Visitors. Those from 1710 to 1830 are in RGO MSS 617 and 618.

RSMS—Royal Society manuscripts.

SMITH, Robert, 1738. *A Compleat System of Opticks* (London).

C—Secondary sources

BAILY, Francis, 1835. *An Account of the Rev^a John Flamsteed* ... (London: Admiralty).

BAILY, Francis, 1835. 'Some Account of the Astronomical Observations made by Dr Edmund Halley, at the Royal Observatory at Greenwich', *Mem. R. astron. Soc.*, VIII, pp. 169–90.

CUDWORTH, W., 1889. *Life and Correspondence of Abraham Sharp*, pp. 318–28.

GILBERT, G. S., 1933. 'Greenwich Observatory, 1675–1714', *The Observatory* LVI, 704, pp. 22–6.

GRANT, R. 1852. *History of Physical Astronomy.*

HOWSE, Derek, 1970–1. *The Tompion Clocks at Greenwich and the dead-beat escapement* (Antiquarian Horology Reprint).

Howse, Derek, 1973. *Guide to the Old Royal Observatory Greenwich* (London: National Maritime Museum).

King, Henry C., 1955. *The History of the Telescope* (London: Charles Griffin & Co. Ltd.).

Laurie, P. S., 1960. *The Old Royal Observatory. A brief history* (London: National Maritime Museum).

Lewis, T., 1890. 'Notes on some historical instruments at the Royal Observatory Greenwich'. *The Observatory*, v 163, pp. 200–5.

McCrea, W. H., 1975. *The Royal Greenwich Observatory: an historical review issued on the occasion of tercentenary* (London: H.M.S.O.).

APPENDIX I
Summary of Building Changes at Greenwich
(An appendix to Chapter 1 illustrated in Appendix 3)

1. *Flamsteed House/Dwelling House*

1675	June, July.	Building set out by Hooke to Wren's design, on foundations of Greenwich Castle.
	August 10	Flamsteed laid foundation stone.
	Christmas	Roof laid.
1676	May 31	First observation in Great Room.
	July 10	Flamsteed and staff moved in.
1721		Halley built shed for transit onto west end.
Between 1750 and 1770		Extension southward. New kitchen and living-room, single tall chimney.
After 1794		Two more rooms built onto southern extension. New front door.
1817		Theodolite placed on roof (removed 1842).
1819		N.W. dome joined to house by Zenith Sector apartment. N.E. dome joined by corresponding wall.
1833		Time-ball on east turret.
1835–6		New rooms on west side, ground floor. Two tall chimneys.
1840		Osler anemometer to west turret.
1866		Robinson anemometer in hut on roof.
1840		Porch and covered way to New Observatory.
1849		External staircase from front court to roof.
1908		New porch and covered way.
1911		Basement rooms excavated under west rooms of 1835–6 on ground floor.
1925		Master clocks to cellars.
1940		Astronomer Royal moved to Abinger for period of war.
1945		AR returned to Greenwich.
1948	August	AR moved to Herstmonceux. Flamsteed House ceased to be AR's official residence.
1953	May 8	Octagon Room opened to public by Duke of Edinburgh.
1958–60		Flamsteed House restored and altered to become part of Museum. External ladder and tall chimneys removed.
1960	July 6	Opened as part of NMM by H.M. The Queen.

2. Terrace Summer Houses/North Domes

1676	Built, probably after the main house was complete. East became solar observatory; west, a bedroom.
1773	Converted into observatories for equatorial sectors: both raised one storey, extended southwards, and fitted with hemispherical domes.
1816	Shuckburgh equatorial mounted in E. dome.
1819	W. dome joined to house by Zenith Sector Apartment. Wall built to join E. dome to house.
1846	W. dome became a bedroom.
1856	Basement of E. dome used for batteries for electric clocks. Chronograph installed in ground floor.
1915	Chronograph to East Building.
1927	Rugby time-signal apparatus in cellar of E. dome.
1929	Shuckburgh equatorial to Science Museum.
1938	Quartz crystal clock installed in Battery Basement.

3. Flamsteed's Observatory/Advanced Building/South Dome

1675–6	Built as Quadrant House and Sextant House.
1721	Converted into pigeon house by Halley.
1762	Fitted with hemispherical dome for movable quadrant.
1779	Dome replaced by flat inclined roof with sliding shutters, for small telescopes. Joined to Quadrant Room; known henceforward as Advanced Building.
1844	New South Dome for altazimuth raised on walls of Advanced Building. Drum dome.
1911	Altazimuth replaced by Photoheliograph. Same dome.
1949	Photoheliograph to Herstmonceux. Dome removed and upper parts demolished.
1965–7	Restored to 1695 state and opened to public.

4. Bradley's New Observatory/Meridian Building West

1725	Halley erected quadrant wall with building over.
1749	Halley's building demolished. New Observatory built, comprising Quadrant Room, Transit Room, Middle Room (Library) with Assistant's bedroom above.
1779	Hip roof removed and ridge extended to gable. Flat ceilings replaced by double roof. Shutter openings widened from 6 inches to 3 feet.
1809	New Circle Room built onto east end. Gothic window.
1813	East Building built onto east end of Circle Room.
1819	Central heating installed in observing rooms.
1835	Computing Room extended southwards a few feet.

1839–41	Covered way from Dwelling House to Quadrant and Circle Rooms. Old door from Computing Room to Courtyard blocked. Quadrant Room divided into Manuscript Room (west) and Passage Room (east). Computing Room enlarged by moving stairs to Passage Room.
1848–50	Circle Room converted to Transit Circle Room by extending southwards and making roof ridge run north and south. New door to Front Court.
1851	Transit Room became AR's Official Room; Upper Room became Chronometer Room.
1855	Reflex Zenith Tube built south of AR's Room.
1869	Chronometer Room became Upper Computing Room. Chronometers to S.E. dome.
1890	Additional Computing Room built over old Quadrant Room, surmounted by Astrographic Dome. Spiral staircase.
1899	AR moved to new building. Old Transit Room became Chronometer Store and watch-rating room.
1911	Manuscript Room (old Quadrant Room west) became Clock Room.
1937	Room under Astrographic dome became Chronometer Workshop.
1956	Astrographic telescope to Herstmonceux. Dome removed.
1965–7	Bradley observatory restored to 1779 state and opened to public with Transit Circle Room. Renamed Meridian Building.

5. East Building/Meridian Building East

1813	Built as accommodation for assistants and library. Dome built for Shuckburgh Telescope but not so used.
1821	Library became Chronometer Room. Extra storey added for Library.
1838	Sheepshanks equatorial installed in dome.
1851	Chronometer Room became extension to Library. Chronometers to Upper Room in Bradley building.
1854–5	Fireproof record rooms built onto east end. Two storeys.
1915	Chronograph from N.E. dome to ground floor.
1963	Sheepshanks equatorial to Altazimuth Pavilion.
1965–7	Converted into John Pond and Frank Dyson Galleries. Opened to public.

6. Front Court

1676	Main gate in wall about 25 feet from house.
1779	New Great Gate.
After 1794	Present Front Court enclosed.
1826	Ramage's telescope erected in courtyard.
1836	Ramage's telescope removed.
1840	Covered way from Flamsteed House to Quadrant and Circle Rooms.
1849	External staircase to roof of Flamsteed House.

1889	Brick Porter's lodge in place of wooden hut.
1891	Transit Pavilion built on Bradley meridian.
1908	New covered way.
1911	Hut for Cookson telescope built west of Transit Pavilion.
1940	Gatehouse demolished by bomb.
1948	New gatehouse.
1956–60	Transit Pavilion, Cookson hut and gatehouse demolished.

7. Great Equatorial Building/South-East Equatorial Building

1857	Erected with wooden drum dome for Great Equatorial telescope. Chronometers on first floor.
1892–3	Drum dome replaced by onion-shaped dome for new 28-inch refractor.
1895	Chronometers spread to ground floor.
1898	Balcony erected around dome.
1908	Railings on shutter platform to form 'crown'.
1940–4	Dome damaged by bombs.
1947	28-inch telescope to Herstmonceux.
1953	Dome removed, building remained.
1968	Ground and first floors taken over as accommodation for museum warders.
1971	28-inch telescope re-mounted.
1974–5	New onion dome erected.

8. South Ground/New Observatory/South Building/etc.

1817	Wooden magnet house erected.
1824	Demolished as dangerous.
1836–7	New cruciform wooden Magnet House erected.
1863	Magnet House basement excavated. New magnetic buildings to south.
1873	Photoheliograph dome erected.
1881	New Library built (later, the Central Store).
1884	Lassell dome erected at ground level.
1891–9	New Physical Observatory built, surmounted by Lassell Dome housing Thompson equatorial.
1895	New stables.
1899	New Altazimuth Pavilion.
1908	New storehouse.
1917	Magnetic Observatory demolished.
1949	Thompson equatorial to Hertsmonceux.
1963	Sheepshanks telescope to Altazimuth Pavilion.
1965	Planetarium installed in Thompson Dome. Picture restoration studios in ground floor and basement.

9. Christie Enclosure

1899	Enclosed. Magnetic Pavilion built.
1931	Magnetic Pavilion demolished. Buildings for Yapp telescope, Reversible Transit Circle and Cookson telescope erected.

| 1953 | RTC to Herstmonceux. |
| 1956–7 | All buildings demolished. Enclosure returned to park. |

10. *Grounds and Gardens*

1676	All gardens within perimeter. Mast in garden.
c. 1690	Mast removed.
1748	Middle Garden enclosed with New Observatory.
c. 1770	Garden House and stables built under wall of Upper Garden.
After 1794	Front Court enclosed.
1813	New wall for 'drying ground' near new East Building.
1814	Lower Garden (vegetables) enclosed.
1836	South ground enclosed.
1890	South ground extended.
1899	Christie enclosure made.
1957	Christie enclosure returned to park.

APPENDIX II
*The Principal Magnetic and Meteorological Instruments used at Greenwich
and Abinger*
(An appendix to Chapter 10)

Abbreviations

Chris. Encl.	= Christie enclosure, Greenwich
Mag. Ho. (U)	= Upper Magnet Room, Greenwich
Mag. Ho. (L)	= Magnetic Basement, Greenwich
Mag. Gd.	= Magnetic Ground (South Ground), Greenwich
Mag. Pav.	= Magnetic Pavilion, Christie Enclosure
M'graph Ho.	= Magnetograph House, Christie Enclosure
NMM	= National Maritime Museum, Greenwich
NPL	= National Physical Laboratory, Teddington
Oct. Rm.	= Octagon Room, Greenwich
RTC	= Reversible transit circle
Sc.M.	= Science Museum, London

1817	Pond's Magnet House built.
1824	Same abandoned.
1838	Airy's Magnet House built.
1840	Magnetical and Meteorological Department formed under James Glaisher. Regular observations started November.
1847	Photographic registration commenced (wet paper).
1863	Magnetic Basement excavated, new magnetic huts built.
1882	Dry paper photographic registration introduced.
1899	Absolute intensity and dip instruments transferred to new Magnetic Pavilion in Christie Enclosure, with some meteorological instruments.
1914	Variation instruments transferred to new Magnetograph House in Christie Enclosure.
1918	Airy's Magnet House demolished.
1925–6	Magnetic observations transferred to Abinger.

154

1932	Buildings for telescopes (Yapp, RTC and Cookson) erected in Christie Enclosure.	
1950	Meteorological observations transferred from Greenwich to Herstmonceux.	
1957	Magnetic observations transferred from Abinger to Hartland Point.	

Magnetic Instruments

Magnetic elements measured:

> D = Magnetic declination (or variation)
> I = Magnetic inclination (or dip)
> H = Horizontal intensity
> Z = Vertical intensity

Date in use —and where	Description	Present whereabouts
Absolute measurement of D		
1680–1716	Flamsteed uses Royal Society's 1-foot needle, Collins's 8-inch and other needles.	
1749–57	Bradley's horizontal needle. No details.	
1817–19	Pond's needle by Dollond, with two reading microscopes.	
1838–1900 Mag. Ho. (U)	Upper declination magnet by Meyerstein of Göttingen, 2-foot magnet, 8-foot 9-inch suspension. Used also for variation of D, 1847–63.	NMM (MI.3) ex. Sc.M (Inv. 1925–954)
1899–1925 Mag. Pav.	Declinometer by Elliot, No. 75, with 4-inch hollow cylindrical magnet and tungsten suspension.	Herstmonceux
1925–57 Abinger	Similar declinometer by Elliot, used in conjunction with Watts theodolite.	Hartland
Absolute measurement of Dip		
1749–57	Bradley's dip instrument.	
1817–19	Pond's dipping needle (possibly inherited from Maskelyne or even Bradley).	
1843–61 Dip hut	9-inch dip circle by Robinson.	
1861–3 Dip hut 1863–83 Dip office	Airy's dip circle, signed: *Troughton & Simms London 1860*, 3-inch, 6-inch, and 9-inch needles. (Fig. 115).	NMM (MI.1.) ex Sc.M (Inv. 1925–957)

Dates in use —and where	Description	Present whereabouts
1883–98 New library		
1898–1915 Mag. Pav.		
1914–26 Mag. Pav	Dip inductor by Cambridge Instru- ment Co. (Fig. 116)	Herstmonceux
1925–52 Abinger	Similar instrument by Cambridge Inst. Co.	Hartland
1929–57 Abinger	Dye coil magnetometer (see Absolute Z below) and Schuster Smith coil magnetometer (see Absolute H below) used to compute dip.	Hartland

Absolute measurement of H

1840–61 Mag. Ho. (U)	Gauss bifilar horizontal force mag- netometer by Meyerstein of Göt- tingen, 2-foot magnet, 7-foot 9-inch suspension. Later history under Variation of H.	NMM (MI.6) ex Sc.M. (Inv. 1925–955)
1861–99 Deflexion hut 1899–1926 Mag. Pav.	Kew-pattern unifilar magnetometer (used as deflexion instrument) by Gibson, No. 3. (Fig. 117)	
1925–57 Abinger	Kew-pattern unifilar magnetometer by Casella, No. 181—used as check on coil magnetometer.	Hartland
1925 Mag. Pav. 1927–57 Abinger	Schuster Smith coil magnetometer (lent by NPL).	Hartland

Absolute measurement of Z

Until 1929, absolute Z was deduced from H and dip.

1929–57 Abinger	Dye vertical-force coil magnetometer (lent by NPL).	Hartland

Variations of D

1847–64 Mag. Ho. (U)	Upper declination magnet by Meyerstein used for both absolute and variation measurements.	NMM (MI.3) ex Sc.M. (Inv. 1925–954)
1864–1914 Mag. Ho. (L)	Declination magnetograph by Troughton & Simms, 2-foot magnet, 6-foot suspension.	NMM (MI.6) ex Sc.M. (Inv. 1925–955)

Dates in use —and where	Description	Present whereabouts
1914–26 M'graph Ho.	Declination variometer by Cambridge Instrument Co., 4·5-cm magnet, 30-cm phosphor-bronze suspension.	
1925–38 Abinger	Declination variometer, 10-mm needle, quartz-fibre suspension.	
1938–57 Abinger	La Cour variometer, 8-mm magnet, quartz-fibre suspension.	Hartland

Variations of H

1847–64 Mag. Ho. (U)	Gauss bifilar magnetometer by Meyerstein, used for both absolute	NMM (MI.6) Sc.M.
1864–1914 Mag. Ho. (L)	and variation measurements 1847–64	(Inv. 1925–955)
1914–26 M'graph Ho.	North-force variometer by Cambridge Instrument Co., 4·5-cm needle, 20 cm quartz suspension.	
1925–38 Abinger	Horizontal-force variometer, 10 mm needle, quartz-fibre suspension.	
1938–57 Abinger	La Cour variometer, 8-mm magnet, quartz-fibre suspension.	Hartland

Variations of Z

1847–64 Mag. Ho. (U)	Balance magnetometer, 2-foot magnet by Barrow on agate planes.	
1864–1917 Mag. Ho. (L)	Vertical-force magnetograph, 18-inch magnet with pointed ends, by Simms	NMM (MI.4) ex Sc.M. (Inv. 1925–952)
1915–21 M'graph. Ho.	Watson vertical-force variometer, 8-cm magnets, quartz thread suspension (on loan from Met Office).	
1921–6 M'graph Ho.	New Watson quartz-thread variometer, by Hilger.	
1926–38 Abinger	Similar Watson quartz-thread variometer.	
1938–57 Abinger	La Cour variometer, 61 mm magnet, knife edges on agate.	Hartland

Theodolite for absolute measurement of D and variations of H and Z

1838–99 Mag. Ho. (U)	Theodolite by Simms, 8·3-inch horizontal circle, probably acquired	NMM
1899–1926 Mag. Pav.	by Pond *c.* 1818, used with chronometer Parkinson & Frodsham 3719.	
1925–57 Abinger	Theodolite by Watts.	

157

Dates in use —and where	Description	Present whereabouts
	Meteorological Instruments	

Atmospheric pressure

1676–82	Flamsteed's 'baroscope' mentioned in his letters to Richard Towneley, with readings.	
1748 and later	Two barometers by Bird. One reported destroyed 1838, one in Circle Room 1840 and still on inventory 1929.	
1772	Two more barometers. One by Nairne survives.	NMM (MT/BM.24)
1840–1916 Mag. Ho. (U) 1917–1950 M'graph Ho. 1950 RGO Met. hut	Standard barometer signed: *I. Newman 122 Regent St. London No. 64 Diameter of Tube ·565*	Herstmonceux
1848–63 Mag. Ho. (U) 1863–1916 Mag. Ho. (L) 1916–29 M'graph Ho.	Photographic recording barometer.	
1865–1952 Outside gate 1967 Pond gall.	Public barometer, max. and min.	NMM

Temperature, humidity, etc.

1748	Two thermometers by Bird.	
1836–40 Circle Room 1840–76 Mag. Ho.	Standard dry bulb thermometer, property of James Glaisher.	
1840	Daniell's hygrometer.	
1848–87 Mag. Gd.	Photographic self-registering wet and dry thermometers.	
1876–19	Kew Committee standard dry bulb thermometer No. 515.	
1887–1916 Mag. Gd. 1916–38 Chris. Encl.	Christie's photographic self-registering wet and dry bulb thermometers by Negretti & Zambra.	
1938–57 Chris. Encl.	Distant-reading thermograph.	

N.B. Many other thermometers are mentioned in the records.

Dates in use —and where	Description	Present whereabouts
Wind speed and direction		
1840–1953 Oct. Rm. roof (N.W. turret)	Osler self-registering anemometer by Newman. Incorporated pluvio- meter (see rain-gauges below). (Figs. 118 and 120).	
1843–62 Oct. Rm. roof	Whewell's self-registering anemometer by Simms.	Sc.M. (Inv. 1893–166)
1859–66 Oct. Rm. roof	Robinson's anemometer by Negretti & Zambra.	
1866–1950 Oct. Rm. roof	Robinson's anemometer by Browning (Fig. 119).	
Rainfall		
1676–82	Flamsteed's rain gauge, mentioned in letters to Sir Jonas Moore.	
c. 1820–c. 1900	Pond's rain gauge by Troughton.	NMM (MT/R.1)
1840–1953 Oct. Rm. roof (N.W. turret)	Pluviometer attached to Osler's anemometer. (Fig. 118).	

N.B. Many other rain gauges are mentioned in the records.

Sunshine and clouds		
1840	Actinometer for measuring intensity of solar radiation.	
1876–87 Mag. Gd.	Campbell sunshine recorder.	
1887–96 Mag. Gd.	Campbell-Stokes sunshine recorder (numbered MO 113 in 1926).	
1896–1951 Oct. Rm. roof		
1904	Fineman nephoscope.	
1920–50 Front Court	Polar night-sky camera, quarter plate, 48 cm focal length.	

Atmospheric electricity		
1842–78 Mag. Ho. (U)	Several electrometers connected to 79-foot pole.	
1878–1917 Mag. Ho. (U)	Thomson self-recording quadrant electrometer by Whites of Glasgow.	
1917–31 Electrometer hut		

Earth currents		
1865–90 Mag. Ho. (L)	Earth current galvanometer. Dis- continued due to advent of electric trains.	NMM (MI.2) ex Sc.M. (Inv. 1925–1016)

Dates in use —and where	Description	Present whereabouts
Ozone 1876–1901 Mag. Gd. 1901–10 Chris. Encl.	Ozonometer by Horne & Thornthwaite.	

References

GLAISHER, James, 1844. 'The Magnetic and Meteorological Royal Observatory, Greenwich', *The Illustrated London News*, 16 March, 1844, pp. 163–4.

GLAISHER, James, and DUNKIN, Edwin, 1862, pp. 42–3. (See Bibliography.)

MALIN, Stuart, 1974. 'Table of Declination and Dip measurements made in London' (in preparation).

MAUNDER, E. Walter, 1900, pp. 228–50. (See Bibliography.)

ROYAL OBSERVATORY, annually from 1840 to 1957. *Results of Magnetical and Meteorological Observations.*

The Growth of the Royal Observatory, Greenwich

1675–1720

1720–1765

1765–1835

1835–1881

1881–1910

Christie Enclosure 1899–c. 1956–7

In each phase, new buildings are shown hatched.

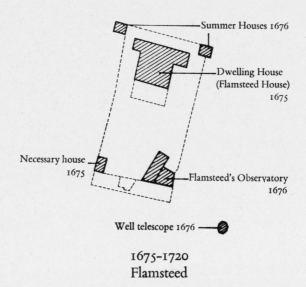

Summer Houses 1676

Dwelling House
(Flamsteed House)
1675

Necessary house
1675

Flamsteed's Observatory
1676

Well telescope 1676

1675-1720
Flamsteed

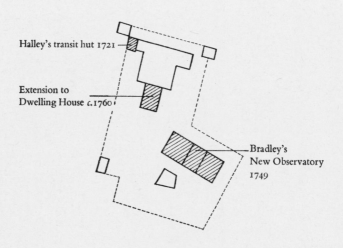

Halley's transit hut 1721

Extension to
Dwelling House c.1760

Bradley's
New Observatory
1749

1720-1765
Halley, Bradley, Bliss

162

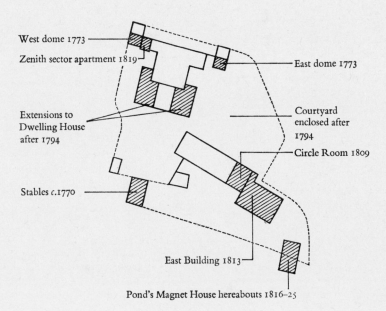

West dome 1773

Zenith sector apartment 1819

East dome 1773

Extensions to
Dwelling House
after 1794

Courtyard
enclosed after
1794

Circle Room 1809

Stables c.1770

East Building 1813

Pond's Magnet House hereabouts 1816–25

1765–1835
Maskelyne, Pond

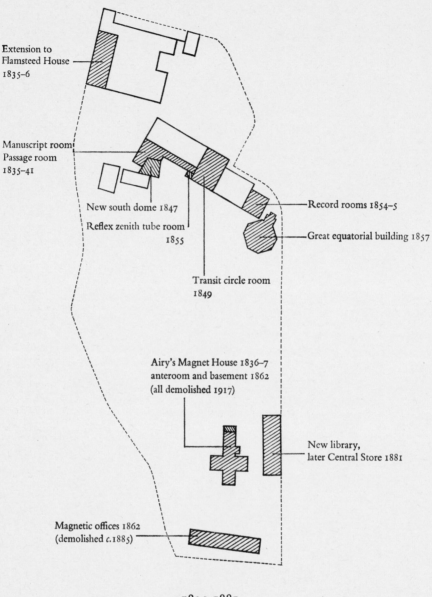

Extension to
Flamsteed House
1835–6

Manuscript room
Passage room
1835–41

New south dome 1847

Reflex zenith tube room
1855

Transit circle room
1849

Record rooms 1854–5

Great equatorial building 1857

Airy's Magnet House 1836–7
anteroom and basement 1862
(all demolished 1917)

New library,
later Central Store 1881

Magnetic offices 1862
(demolished c.1885)

1835–1881
Airy

164

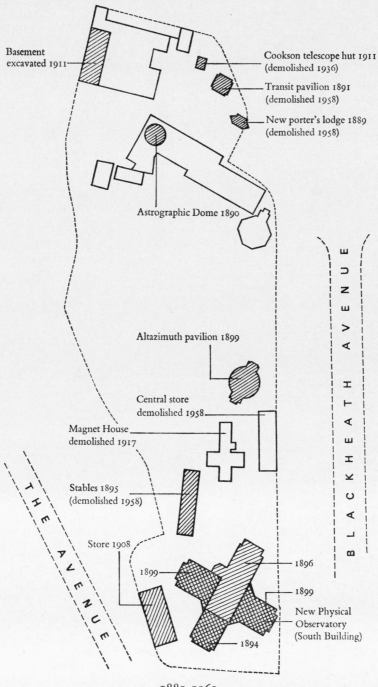

Basement
excavated 1911

Cookson telescope hut 1911
(demolished 1936)

Transit pavilion 1891
(demolished 1958)

New porter's lodge 1889
(demolished 1958)

Astrographic Dome 1890

Altazimuth pavilion 1899

Central store
demolished 1958

Magnet House
demolished 1917

Stables 1895
(demolished 1958)

Store 1908

1899

1896

1899

New Physical
Observatory
(South Building)

1894

THE AVENUE

BLACKHEATH AVENUE

1881–1960
Christie, Dyson, Spencer-Jones, Woolley

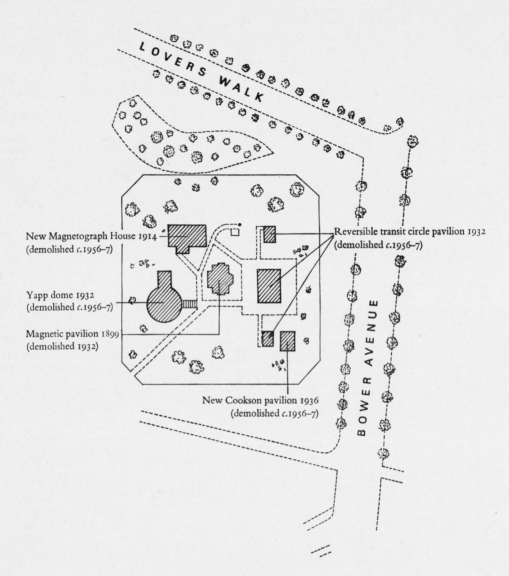

New Magnetograph House 1914
(demolished c.1956–7)

Yapp dome 1932
(demolished c.1956–7)

Magnetic pavilion 1899
(demolished 1932)

Reversible transit circle pavilion 1932
(demolished c.1956–7)

New Cookson pavilion 1936
(demolished c.1956–7)

LOVERS WALK

BOWER AVENUE

1899–c. 1956–7
Christie Enclosure (Magnetic Enclosure)
350 yd east of Flamsteed House

Index

1. The principal references are given in Bold type.
2. Until about 1830, the size of a telescope was generally indicated by quoting its focal length rather than the diameter of the objective, as is the current practice. In this book, we have used whichever convention was in favour at the time the telescope concerned was operational.
3. Metric equivalents are given in the index but not in the text.